T0296201

Cambridge Tracts in Mathematics and Mathematical Physics

General Editors

J. G. LEATHEM, M.A.
E. T. WHITTAKER, M.A., F.R.S.

No. 8

The Elementary Theory of the Symmetrical Optical Instrument

THE ELEMENTARY THEORY OF THE SYMMETRICAL OPTICAL INSTRUMENT

by

J. G. LEATHEM, M.A.
Fellow and Lecturer of St John's College
and University Lecturer in Mathematics

CAMBRIDGE:
at the University Press
1908

CAMBRIDGE
UNIVERSITY PRESS

University Printing House, Cambridge CB2 8BS, United Kingdom

Cambridge University Press is part of the University of Cambridge.

It furthers the University's mission by disseminating knowledge in the pursuit of education, learning and research at the highest international levels of excellence.

www.cambridge.org
Information on this title: www.cambridge.org/9781107493599

First published 1908
Re-issued 2015

A catalogue record for this publication is available from the British Library

ISBN 978-1-107-49359-9 Paperback

PREFACE

IN Gauss's *Dioptrische Untersuchungen* there is little trouble with sign conventions, and continued fractions are not employed. These are, however, prominent features in more recent presentations of the first-order theory of the optical instrument, and render the subject somewhat difficult to the beginner.

It is the aim of the present Tract to eliminate all unnecessary difficulties and to give a quite elementary account of the theory; and, to this end, it has seemed desirable to follow (in Sections I—IV) the general lines of Gauss's memoir. I am indebted to a suggestion of Mr T. J. I'A. Bromwich for a feature of the present scheme which seems to me to mark a great advance in simplification, namely the postponing of the study of the functional form of the constants of the instrument till after its general optical properties have been established, and the employing of an elementary theorem in algebraic linear transformations to obtain the fundamental equations and the relation between the constants.

Limits of space have prevented any close examination of the application of the theory to particular instruments, but one or two questions in connection with the equivalent thin lens and the adjustment of field-glasses, not usually treated in text-books, have been discussed; and a few pages have been devoted to bringing reflecting instruments within the scope of the theory.

In Section IX a brief and, I hope, easy account is given of the third-order aberrations.

I am deeply indebted to Mr Bromwich for reading the manuscript, for his assistance in drawing up the syllabus which constitutes Section X, and for other most valuable suggestions. My thanks are due to Mr W. M. Page, Fellow of King's College, for reading the proofs, and to Mr S. D. Chalmers, of the Northampton Institute, for giving me the benefit of his knowledge of technical optics.

J. G. L.

ST JOHN'S COLLEGE,
1 *May* 1908.

CONTENTS

PAGE

PREFACE v

I. Approximate formulae for a succession of refractions at nearly normal incidence 1

II. The mathematical solution of the refraction problem for a symmetrical instrument 8

III. The optical properties of a symmetrical instrument . . 17

IV. The manner in which the optical properties of an instrument depend on its constitution 26

V. The equivalent thin lens 36

VI. Reflecting instruments 39

VII. Entrance and exit pupils 43

VIII. Chromatic defects of the image 44

IX. The aberrations of the third order 58

X. Syllabus of propositions concerning the characteristic function and the focal lines of a pencil of rays of light . . . 66

THE ELEMENTARY THEORY OF THE SYMMETRICAL OPTICAL INSTRUMENT.

I. APPROXIMATE FORMULAE FOR A SUCCESSION OF REFRACTIONS AT NEARLY NORMAL INCIDENCE.

1. Analytical formulae expressing the laws of refraction.

These laws are: (i) The incident ray, the refracted ray, and the normal to the refracting surface at the point of incidence are in one plane. (ii) The ratio of the sines of the angles of incidence and refraction is a constant, depending only on the nature of the media in which the light is propagated.

It is important to express these laws in terms of the direction cosines of the lines involved. Let μ be the index of refraction of the medium in which is the incident ray, μ' the index of the medium in which is the refracted ray; let the cosines of the incident ray be (l, m, n), those of the refracted ray (l', m', n'), those of the normal to the refracting surface at the point of incidence, drawn towards the medium μ', (L, M, N); let the angles of incidence and refraction be ϕ and ϕ' respectively.

In the diagram AO is the incident ray, OC the refracted ray, and OK the normal. On the incident ray produced a point B is taken so that OB is μ units of length, and on the refracted ray a point C is taken so that OC is μ' units of length. BJ, CK are perpendiculars drawn to the normal. Then the angles BOJ, COK are ϕ, ϕ' respectively, and the lengths of JB and KC are $\mu \sin \phi$ units and $\mu' \sin \phi'$ units respectively.

The first law tells us that JB and KC are in the same plane through JK, and therefore parallel to one another. The second law tells us that JB and KC are of the same length. The two laws are expressed in the statement that JB and KC have equal projections on each of the coordinate axes. Now the projection of JB is the excess of the projection of OB over that of OJ; and OB is of length μ and

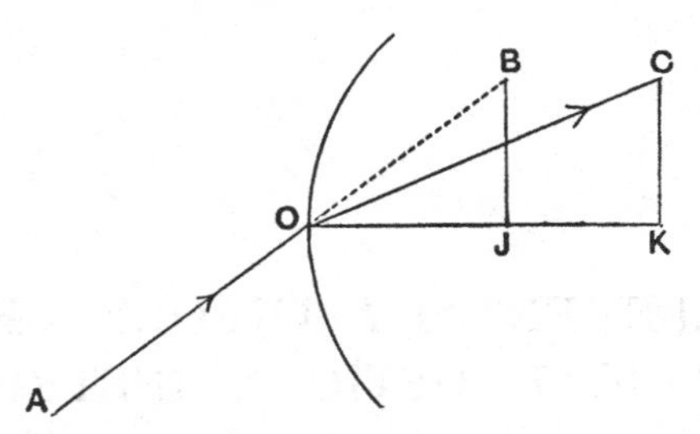

has cosines (l, m, n), while OJ is of length $\mu \cos \phi$ and has cosines (L, M, N). Likewise the projection of KC is the difference of the projections of OC and OK. So the equalities of the projections of JB and KC on the three axes of coordinates are expressed by equations of the type

$$\mu l - \mu \cos \phi L = \mu' l' - \mu' \cos \phi' L.$$

Rearranging, we get the following equations to express the laws of refraction:

$$\left.\begin{aligned} \mu' l' - \mu l &= (\mu' \cos \phi' - \mu \cos \phi) L \\ \mu' m' - \mu m &= (\mu' \cos \phi' - \mu \cos \phi) M \\ \mu' n' - \mu n &= (\mu' \cos \phi' - \mu \cos \phi) N \end{aligned}\right\} \qquad (1).$$

The three equations are not all independent, as is readily seen by multiplying them by L, M, N respectively and adding, it being remembered that $\Sigma L l' = \cos \phi'$, and $\Sigma L l = \cos \phi$.

2. Approximate formulae for nearly normal incidence.

When the incident ray is very nearly normal, it is readily seen that the refracted ray is also very nearly normal. It is therefore possible so to choose the axis of z that it shall be nearly parallel to the incident ray, the refracted ray, and the normal at the point of incidence. When the axis of z has been so chosen, l, m, l', m', L, M, are all small. We shall obtain approximate formulae on the hypothesis that these quantities are so small that their squares and products may be neglected; this is equivalent to the supposition that aberration is to be neglected.

On this hypothesis n, which is equal to

$$(1 - l^2 - m^2)^{\frac{1}{2}},$$

differs from unity by small quantities of the second order; so also

n' and N. Hence we may replace all three by unity. Also ϕ and ϕ' are small of the same order as l, etc.; so $\cos\phi$, $\cos\phi'$ differ from unity by small quantities of the second order, and may be replaced by unity.

Thus the third of equations (1) becomes, on the basis of the proposed approximation, an identity; and the other two take the forms

$$\left.\begin{aligned}\mu' l' - \mu l &= (\mu' - \mu) L \\ \mu' m' - \mu m &= (\mu' - \mu) M\end{aligned}\right\} \quad \text{(2)}.$$

3. Expression of L, M in terms of the coordinates of the point of incidence.

The most usual application of the formulae (2) is to the case in which a pencil of rays pass nearly normally through a comparatively small portion of the refracting surface, so that the rays and the normals at all the points of incidence are very nearly parallel to one another. It is then advantageous to take as axis of z the normal at some one of the points of incidence, say at a point so centrally situated that it may be called the centre of the portion of the surface through which refraction takes place; the ray incident at this point may be called the central ray of the pencil; it need not be more precisely defined.

Unless the point at which the axis of z is normal, say $(0, 0, c)$, is a singular point on the refracting surface, the part of the surface in the neighbourhood of this point can be approximately represented by an equation of the type

$$2(z - c) + ax^2 + 2hxy + by^2 = 0 \quad \text{(3)},$$

the approximation neglecting terms of the third order in x and y. The constants a, h, b are of course known when the shape of the surface is known.

At the point (x, y, z) the cosines of the normal are approximately

$$(ax + hy),\ (hx + by),\ 1,$$

when the squares and products of x, y are neglected. The first two of these may be substituted for L, M in the formulae (2), and so we get the refraction of the ray incident at (x, y, z) determined by the equations

$$\left.\begin{aligned}\mu' l' - \mu l &= (\mu' - \mu)(ax + hy) \\ \mu' m' - \mu m &= (\mu' - \mu)(hx + by)\end{aligned}\right\} \quad \text{(4)}.$$

It is, of course, possible to choose the coordinate planes of x and y

so that the h of formula (3) shall be zero. If this were done the approximate equation of the surface would take the form

$$2(z-c)+\frac{x^2}{\rho_1}+\frac{y^2}{\rho_2}=0 \quad \text{.....................(5)},$$

and ρ_1, ρ_2 would be the principal radii of curvature of the surface at the point $(0, 0, c)$, reckoned positive when the corresponding convexities of the surface are towards the medium μ'. The formulae which would then take the place of (4) are

$$\left.\begin{aligned} \mu' l' - \mu l &= \frac{\mu'-\mu}{\rho_1}x \\ \mu' m' - \mu m &= \frac{\mu'-\mu}{\rho_2}y \end{aligned}\right\} \quad \text{..................(6)}.$$

4. A series of refracting surfaces having a common normal.

When a ray traverses a succession of different media arranged in such a way that the refracting surfaces have a common normal with which the ray is always nearly coincident, it is interesting to see how the equations of the previous Article enable us to derive from a knowledge of the position of the ray before its first incidence a complete specification of the ray after its final emergence.

The incident ray (say in a medium μ_0) is known when we know its cosines l_0, m_0, and the coordinates (x_1, y_1, z_1) of the point where it meets the first surface,

$$2(z-c_1)+a_1x^2+2h_1xy+b_1y^2=0.$$

Clearly z_1 differs from c_1 only by quantities of the second order, so we may replace z_1 by c_1, and regard the incident ray as specified by the four quantities l_0, m_0, x_1, y_1.

After the first refraction (into a medium μ_1) the cosines of the ray are changed to l_1, m_1, given by the equations

$$\mu_1 l_1 - \mu_0 l_0 = (\mu_1-\mu_0)(a_1x_1+h_1y_1),$$

$$\mu_1 m_1 - \mu_0 m_0 = (\mu_1-\mu_0)(h_1x_1+b_1y_1).$$

Equations of this type, which determine the change of direction due to refraction, may be called "optical" equations.

The coordinates (x_2, y_2, c_2) of the point where the refracted ray meets the second refracting surface

$$2(z-c_2)+a_2x^2+2h_2xy+b_2y^2=0,$$

can be obtained by putting $z=c_2$ (a sufficient approximation) in the equations of the ray, i.e. of the line which proceeds from (x_1, y_1, c_1) in the direction $(l_1, m_1, 1)$. Now the equations of this line are

$$(x - x_1)/l_1 = (y - y_1)/m_1 = z - c_1,$$

and so x_2, y_2 are determined by the equations

$$\left.\begin{aligned} x_2 &= x_1 + l_1\,(c_2 - c_1) \\ y_2 &= y_1 + m_1(c_2 - c_1) \end{aligned}\right\} \text{.......................(7).}$$

Equations of this type, which determine the coordinates of a point of refraction in terms of those of the previous point of refraction, may be called "geometrical" equations.

Having found x_2, y_2 by the geometrical equations, we are in a position to use the optical equations corresponding to the next refraction, viz.

$$\mu_2 l_2 - \mu_1 l_1 = (\mu_2 - \mu_1)\,(a_2 x_2 + h_2 y_2),$$
$$\mu_2 m_2 - \mu_1 m_1 = (\mu_2 - \mu_1)\,(h_2 x_2 + b_2 y_2),$$

equations which we could not use till we knew x_2, y_2, but which now give us the values of l_2, m_2.

Eliminating l_1, m_1 from the six equations we are left with four equations which give us explicit formulae for x_2, y_2, l_2, m_2 in terms of l_0, m_0, x_1, y_1; that is, the quantities specifying the ray in the medium μ_2 in terms of the quantities that specify the incident ray.

Thus the problem of refraction is solved for the case of two surfaces. If there are n surfaces and $n+1$ media, we get in a similar manner $2n$ optical equations and $2n-2$ geometrical equations. From these we can eliminate successively the $4n-6$ quantities

$$l_1,\ m_1,\ x_2,\ y_2,\ l_2,\ m_2,\ x_3,\ y_3,\ \ldots\ x_{n-1},\ y_{n-1},\ l_{n-1},\ m_{n-1},$$

and obtain finally four equations expressing x_n, y_n, l_n, m_n, the quantities specifying the emergent ray, in terms of l_0, m_0, x_1, y_1, the quantities that specify the incident ray*.

5. Case in which all the refracting surfaces are symmetrical about the same two planes through the axis.

By suitable choice of the planes of x and y it is always possible to make the coefficient of xy in the equation of one of the refracting surfaces vanish; but in general this choice would leave the corresponding coefficients for all the other surfaces different from zero. But if the surfaces are such that their indicatrices at the points where they are met by the common normal (the axis of z) all have their principal

* Cf. Prof. R. A. Sampson on Gauss's *Dioptrische Untersuchungen*, *Proc. London Math. Soc.* XXXIX. 1898, p. 33.

axes in the same two directions, then the same choice of coordinate planes will make all the h's vanish simultaneously.

In this case the equations of the preceding Article are greatly simplified, for when all the h's are zero, the equations divide themselves into two sets, one set involving only x's and l's, the other set involving only y's and m's. To solve the problem of n surfaces we now have only to eliminate the $2n-3$ quantities $l_1, x_2, l_2, \ldots x_{n-1}, l_{n-1}$, from the n optical and the $n-1$ geometrical equations of the first set. The result, with suitable change of symbols, serves also for the corresponding equations of the second set; so that in this particular case the labour of elimination is much less than half that required in the general case.

6. Form of the results in the general case.

Taking the x's, y's, l's, m's in the order which presents itself naturally as one follows the course of a ray, we see that each of these quantities is a homogeneous linear function of those that precede it. Consequently when the elimination has been performed we get x_n, y_n, l_n, m_n as homogeneous linear functions of l_0, m_0, x_1, y_1. A more useful result is arrived at if we specify the positions of the entering and the emergent rays, not by the coordinates (x_1, y_1), (x_n, y_n) of the points where they meet the planes $z=c_1$, $z=c_n$, but by the coordinates (ξ_0, η_0), (ξ, η) of the points where they respectively meet two other arbitrarily selected planes $z=c_1-p$, $z=c_n+q$, which we call planes of reference. This implies the introduction of four more geometrical equations, namely two of the type

$$x_1 = \xi_0 + pl_0,$$

and two of the type
$$\xi = x_n + ql,$$

and the addition of x_1, y_1, x_n, y_n to the quantities to be eliminated.

If we denote by a's with double suffix the coefficients in the expressions for l, m, x_n, y_n in terms of x_1, y_1, l_0, m_0, so that, for example, $l = a_{11}x_1 + a_{12}y_1 + a_{13}l_0 + a_{14}m_0$, it is readily verified that the suggested elimination leads to

$$\left.\begin{aligned} \xi &= (a_{31}+qa_{11})\,\xi_0 + (a_{32}+qa_{12})\,\eta_0 \\ &\qquad + (a_{33}+pa_{31}+qa_{13}+pqa_{11})\,l_0 + (a_{34}+pa_{32}+qa_{14}+pqa_{12})\,m_0 \\ \eta &= (a_{41}+qa_{21})\,\xi_0 + (a_{42}+qa_{22})\,\eta_0 \\ &\qquad + (a_{43}+pa_{41}+qa_{23}+pqa_{21})\,l_0 + (a_{44}+pa_{42}+qa_{24}+pqa_{22})\,m_0 \end{aligned}\right\} \ldots(8),$$

and two others which we may regard as contained in these two, since they may be derived by differentiating with respect to q and remembering that

$$\frac{\partial \xi}{\partial q} = l, \quad \frac{\partial \eta}{\partial q} = m.$$

Thus the whole theory of the set of n surfaces is contained (to the degree of approximation postulated at the outset) in two linear equations involving a rather large array of constants*.

7. Optical interpretation of the equations for the general case.

Without going into detail, it may be useful to indicate how these equations may be interpreted optically, in other words to shew how they give us information as to the image formed by the rays of light that proceeded originally from a given bright point. We regard $(\xi_0, \eta_0, c_1 - p)$ as the coordinates of the bright point, and suppose a white screen to be placed in the plane $z = c_n + q$. If we fix attention on the ray which sets out from the source in a given direction, specified by given values of l_0 and m_0, the equations tell us the co-ordinates of the point in which the ray after refraction strikes the screen. If the pencil of refracted rays form a point image, it must be possible, namely by placing the screen where it will receive the image, to make all the refracted rays strike the screen in the same point; in fact, it must be possible to give such a value to q that ξ and η shall be independent of l_0 and m_0. This means that the giving of a suitable value to q makes four coefficients vanish. Though these four conditions are not all independent, they are in general equivalent to three independent conditions, and so cannot be simultaneously satisfied except in special cases.

Failing to obtain a point image, we next try to find the positions of the focal lines of the pencil of emergent rays. For the points in which the rays strike the screen to lie in a straight line, the necessary and sufficient condition is that there should be a linear relation between ξ and η with coefficients independent of l_0 and m_0. This is the case if the same value of λ makes the coefficients both of l_0 and of m_0 vanish in the expression for $\xi + \lambda\eta$. The condition is therefore the vanishing of

$$\begin{vmatrix} a_{33} + pa_{31} + qa_{13} + pqa_{11}, & a_{34} + pa_{32} + qa_{14} + pqa_{12} \\ a_{43} + pa_{41} + qa_{23} + pqa_{21}, & a_{44} + pa_{42} + qa_{24} + pqa_{22} \end{vmatrix}.$$

There are two values of q which satisfy this condition, and these define the positions of the two focal lines of the emergent pencil.

* Amongst the 16 constants there are 6 relations, which are the conditions for the existence of a Characteristic Function. For the form of the relations see Sampson, l.c. pp. 38, 39; for the connexion with the Characteristic Function see Bromwich, *Proc. London Math. Soc.* XXXI. 1899, p. 8.

8. Case of two planes of symmetry.

When the surfaces are as described in Article 5, the mathematical solution of the refraction problem is much simpler. The equations reduce to

$$\left.\begin{aligned}\xi &= (a_{31} + qa_{11})\,\xi_0 + (a_{33} + pa_{31} + qa_{13} + pqa_{11})\,l_0\\ \eta &= (a_{42} + qa_{22})\,\eta_0 + (a_{44} + pa_{42} + qa_{24} + pqa_{22})\,m_0\end{aligned}\right\} \quad \text{......(9)},$$

and the optical interpretation is easier than in the general case.

II. THE MATHEMATICAL SOLUTION OF THE REFRACTION PROBLEM FOR A SYMMETRICAL INSTRUMENT.

9. The symmetrical optical instrument.

We proceed now to the detailed discussion of a very particular case of the general arrangement discussed in Article 4, namely the case in which all the refracting surfaces are symmetrical with respect, not merely to two planes, but to every plane through the axis of z. The refracting surfaces are then necessarily surfaces of revolution having the axis of z as common axis of revolution; they are generally spheres, but are sufficiently well represented by approximate equations of the type

$$2\,(z - c) + \frac{x^2 + y^2}{\rho} = 0 \quad \text{....................(10)}.$$

Such an arrangement is called a symmetrical optical instrument, and is the kind of instrument most frequently employed in every-day life.

The approximate theory here developed applies to any symmetrical instrument which is used in such a way that the rays which it transmits are very nearly parallel to the axis, for example the telescope and opera-glass; it is practically useless in the case of wide-angle instruments such as the microscope or portrait-camera.

10. Power of a single spherical refracting surface.

When refraction takes place from a medium of index μ to a medium of index μ' through a surface represented approximately by equation (10), the optical equations are equations (6) of Article 3, simplified by the equality of both ρ_1 and ρ_2 to ρ. They are, in fact,

$$\left.\begin{aligned}\mu' l' - \mu l &= \left(\frac{\mu' - \mu}{\rho}\right) x\\ \mu' m' - \mu m &= \left(\frac{\mu' - \mu}{\rho}\right) y\end{aligned}\right\} \quad \text{..................(11)}.$$

Thus it appears that, to the approximation contemplated, the optical properties of the spherical refracting surface are all contained in the single constant

$$\frac{\mu' - \mu}{\rho} \quad \text{..............................(12).}$$

This constant is called the "diverging power" or simply the "power" of the surface, and is usually denoted by κ.

It is most important that the definition of the power of a surface should be understood precisely, without confusion on account of the conventions of sign which are implicitly involved in the above formula. These conventions are ultimately two, namely (i) that the light passes from the medium μ into the medium μ', and (ii) that ρ is positive when the convexity of the refracting surface is towards the second medium.

Suppose the direction of the ray of light to be reversed. Then μ' becomes the index of the first medium, and μ that of the second, so that $\mu - \mu'$ must take the place of $\mu' - \mu$ in formula (12); but convexity towards the medium μ' is concavity towards the medium μ, which is now the second medium, so that the case required by definition is that in which ρ is negative, and $-\rho$ must take the place of ρ in the denominator. Thus the power of the surface for a ray going from μ' to μ is

$$\frac{\mu - \mu'}{-\rho},$$

that is, the same as before. Thus it appears that the power of a surface is not altered by reversing the ray, and its definition has nothing to do with the sense in which the ray passes.

The second of the conventions is thus the only one that affects the definition of the power, and its bearing on the definition is clearly summed up in the following rule:

The power of a refracting surface is to be regarded as positive when the medium on the convex side of the surface has a greater refractive index than the medium on the concave side; the power is negative when the medium of greater index is on the concave side.

This is only another way of saying that κ is positive if $\mu' > \mu$ and $\rho > 0$, or if $\mu' < \mu$ and $\rho < 0$.

For a plane refracting surface ρ is infinite, and κ is zero.

The optical equations for a surface of power κ are

$$\mu' l' - \mu l = \kappa x, \quad \mu' m' - \mu m = \kappa y \quad \text{..................(13).}$$

11. Reduced projected inclinations of a ray.

The equations of the ray in the medium μ being of the form

$$\frac{x-\alpha}{l}=\frac{y-\beta}{m}=z-\gamma,$$

it follows that the equation of the projection of the ray on the plane of xz is of the form

$$x-\alpha=l\,(z-\gamma).$$

Hence l is the tangent of the angle that this projection of the ray makes with the axis of z, or, to our degree of approximation, the angle itself. In fact l and m are the inclinations to the axis of z of the projections of the ray on the coordinate planes through that axis. We call l and m the "projected inclinations" of the ray.

Putting $\mu l=\delta$, $\mu m=\epsilon$, we may call δ and ϵ the "reduced projected inclinations" of the ray, the word "reduced" in this connexion meaning that the projected inclinations are multiplied by the index of the medium in which the ray is passing.

In terms of this notation, the optical equations for a single surface take the form

$$\delta'-\delta=\kappa x,\quad \epsilon'-\epsilon=\kappa y \qquad (14).$$

12. Divergence produced by a refracting surface.

If the effect of refraction were an increase in the projected inclinations of each ray, a pencil of rays would, after passing through the surface, be more divergent or less convergent than before; the surface would literally produce divergence. But it is to be noticed that the optical equations give information, not as to the changes produced by the refraction in the projected inclinations, but as to the changes produced in the *reduced* projected inclinations; and the power of the surface measures the degree to which, for a given point of incidence, it is capable of increasing the reduced projected inclinations. It is therefore convenient to abandon the literal meaning of the word "divergence," and to apply the term to what is really quite different, namely increase of the reduced projected inclinations. With this understanding, a diverging surface is one whose power is positive, a converging surface is one whose power is negative. A plane surface produces neither convergence nor divergence.

In the case of an instrument consisting of several surfaces, if the first and last media are the same there is no difference between the literal and the special meanings of divergence. Thus a double convex lens is a converging instrument, a double concave lens is a diverging instrument.

A thin double concave lens of glass, whose two surfaces have the same curvature, produces divergence; and as this results from the successive refractions at two refracting surfaces of equal power, it may be assumed that each surface separately produces divergence. In fact a surface with air on the concave side and glass on the convex side is a diverging surface or surface of positive power. This fact, which is easy to remember, helps one also to remember the rule given in Article 10, that a surface has positive power when the medium of greater index is on the convex side.

13. Reduced distances.

When reduced projected inclinations are substituted for projected inclinations in the geometrical equations, of the type of equations (7) (the equations of a ray in medium of index μ_1), the forms assumed are

$$x_2 = x_1 + \delta_1 \frac{c_2 - c_1}{\mu_1},$$

$$y_2 = y_1 + \epsilon_1 \frac{c_2 - c_1}{\mu_1}.$$

These will be simplified if we introduce new symbols to represent $(c_2 - c_1)/\mu_1$ and similar expressions which occur in other geometrical equations. Now $c_2 - c_1$ is a distance measured along the axis of the instrument; if we define a_1 by the relation

$$(c_2 - c_1)/\mu_1 = a_1 \qquad \text{.........................(15)},$$

we may call a_1 the corresponding "reduced distance." Reduced distance means distance parallel to the axis or along the ray, divided by the index of the medium in which the ray is passing. In other parts of Optics the phrase "reduced distance" is used in a different sense, so care must be taken not to extend the use of the definition here given beyond its present application.

With the notation of (15) the geometrical equations take the forms

$$\left.\begin{aligned} x_2 &= x_1 + a_1\delta_1 \\ y_2 &= y_1 + a_1\epsilon_1 \end{aligned}\right\} \qquad \text{.......................(16)}.$$

14. Refraction at a single spherical surface.

Let the power of the surface be κ, let the first medium have index μ, the second index μ'. Let it be agreed to specify the incident ray by means of its reduced projected inclinations δ, ϵ, and by the coordinates x, y, of the point where it meets a certain plane of reference; and let the emergent ray be defined by its reduced projected inclinations δ', ϵ', and the coordinates x', y', of the point where it meets a second plane of reference. The planes of reference are perpendicular to the axis; the first is at a *reduced* distance u in front of the refracting surface, the second is at a *reduced* distance u' behind the refracting surface. The corresponding *true* distances are accordingly μu and $\mu' u'$. The reduced distance u is to be measured positively from the surface into the first medium, while u' is to be measured positively from the surface into the second medium. This kind of convention will be adhered to throughout. Later on we shall associate u with the position of an object, u' (or a corresponding symbol) with the position of the image, and it is convenient to remember that u is positive for a real object, and u' positive for a real image.

If the coordinates of the point of incidence be ξ, η, we have three pairs of equations at our disposal. The first are the geometrical equations for the incident ray, typified by

$$\xi = x + u\delta.$$

The second are the optical equations for the refraction, typified by

$$\delta' = \delta + \kappa\xi.$$

The third are the geometrical equations for the refracted ray, typified by

$$x' = \xi + u'\delta'.$$

Eliminating ξ from these we get two equations, namely,

$$\left.\begin{aligned} \delta' &= \kappa x + (\kappa u + 1)\,\delta \\ x' &= (\kappa u' + 1)\,x + (\kappa u u' + u + u')\,\delta \end{aligned}\right\} \quad \text{..........(17)*},$$

which express δ' and x' in terms of x and δ, that is, the quantities specifying the refracted ray in terms of those specifying the incident ray. The corresponding equations expressing ϵ' and y' in terms of y and ϵ need not be written down, for it is clear that they are of the same form and involve precisely the same constants. The four

* The order in which the symbols x, δ, δ', x' appear in these equations and in many later ones may, at first, be thought rather unnatural, but it has been deliberately chosen as representing in a sense the historic or chronological order. In the case of the initial ray we know first the point from which it sets out, and second the direction in which it goes; hence x, y precede δ, ϵ. In the case of the emergent ray we learn first (from the optical equations) its direction, and afterwards (from the geometrical equations) the coordinates of its point of arrival; hence δ', ϵ' precede x', y'.

equations constitute a complete solution of the refraction problem for a single spherical surface.

The form of equations (17) must be carefully noted. They are of the type

$$\left.\begin{aligned} \delta' &= ax + b\delta \\ x' &= cx + d\delta \end{aligned}\right\} \quad\text{.........................(18)},$$

where a, b, c, d are constants depending partly on the power of the surface, and partly on the positions of the planes of reference. And further, since

$$(\kappa u + 1)(\kappa u' + 1) - \kappa(\kappa u u' + u + u') \equiv 1,$$

the constants of (18) are subject to the condition

$$bc - ad = 1 \quad \text{......................... (19).}$$

It will be shewn that the theory of any symmetrical optical instrument, no matter of how many surfaces it is composed, is contained in equations of the same type as (18), with the relation (19) between the constants.

15. Algebraical lemma on linear substitutions.

Let two variables, θ_1, ϕ_1 be homogeneous linear functions of two other variables θ_2, ϕ_2, determined by the relations

$$\theta_1 = p\theta_2 + q\phi_2,$$
$$\phi_1 = r\theta_2 + s\phi_2;$$

and let θ_2, ϕ_2 be similar functions of another pair of variables θ_3, ϕ_3, namely

$$\theta_2 = p'\theta_3 + q'\phi_3,$$
$$\phi_2 = r'\theta_3 + s'\phi_3.$$

From these equations it follows, by the elimination of θ_2 and ϕ_2, that θ_1 and ϕ_1 are expressible as homogeneous linear functions of θ_3, ϕ_3, namely

$$\theta_1 = p''\theta_3 + q''\phi_3,$$
$$\phi_1 = r''\theta_3 + s''\phi_3,$$

where, in fact,

$$p'' = pp' + qr', \quad q'' = pq' + qs',$$
$$r'' = rp' + sr', \quad s'' = rq' + ss'.$$

Now it is clear from these formulae that p'', q'', r'', s'' are the elements, formed by the ordinary rule for multiplying determinants, of the determinant which is the product of

$$\begin{vmatrix} p, & q \\ r, & s \end{vmatrix} \text{ and } \begin{vmatrix} p', & r' \\ q', & s' \end{vmatrix};$$

and therefore

$$\begin{vmatrix} p'', & q'' \\ r'', & s'' \end{vmatrix} = \begin{vmatrix} p, & q \\ r, & s \end{vmatrix} \times \begin{vmatrix} p', & q' \\ r', & s' \end{vmatrix}.$$

That is to say, θ_1 and ϕ_1 are homogeneous linear functions of θ_3 and ϕ_3, and the determinant formed by the coefficients of these functions is equal to the product of the determinants formed respectively by the coefficients of the expressions for θ_1, ϕ_1 in terms of θ_2, ϕ_2, and the expressions for θ_2, ϕ_2 in terms of θ_3, ϕ_3.

The theorem is true equally for three sets of n variables, whatever integral value n may have; but for our purpose the case of $n = 2$ is sufficient*.

If now we consider not merely three, but any number of sets of two variables, the sets being taken in a definite order, and those of each set being homogeneous linear functions of those of the set next in order, we find by repeated application of the theorem just proved that the variables of the first set can be expressed as homogeneous linear functions of those of the last set, and that the determinant of the corresponding coefficients is the product of all the determinants of the coefficients of the relations between intermediate pairs of consecutive sets.

16. Symmetrical instrument consisting of any number n of refracting surfaces.

In the case of an instrument consisting of any number of refracting surfaces, let the indices of the first and last media be μ_0 and μ respectively, and let the powers of the surfaces be $\kappa_1, \kappa_2, \ldots \kappa_n$. Let the entering ray be specified by its reduced projected inclinations δ_0, ϵ_0, and by the coordinates x_0, y_0 of the point where it strikes a plane of reference at reduced distance u in front of the first surface of the instrument. Let the emerging ray be specified by similar quantities δ, ϵ, x, y, and a plane of reference at reduced distance v behind the last surface. Let the ray in any intermediate medium (μ_r) be specified by δ_r, ϵ_r, x_r, y_r, and a plane of reference chosen anywhere in the medium.

Then, by Article 14, it is seen that the δ's and x's of consecutive

* The theorem is a particular case of a well known theorem concerning Jacobians, namely that if θ and ϕ are functions of u and v, and if u and v are functions of x and y, then

$$\frac{\partial(\theta, \phi)}{\partial(x, y)} = \frac{\partial(\theta, \phi)}{\partial(u, v)} \times \frac{\partial(u, v)}{\partial(x, y)}.$$

rays are connected by equations of the same type as (18), subject to a condition of the type of (19). In fact

$$\left.\begin{aligned} x_1 &= c_1 x_0 + d_1\delta_0 \\ \delta_1 &= a_1 x_0 + b_1\delta_0 \end{aligned}\right\}$$

subject to

$$\begin{vmatrix} c_1, & d_1 \\ a_1, & b_1 \end{vmatrix} = 1.$$

So also

$$\left.\begin{aligned} x_2 &= c_2 x_1 + d_2\delta_1 \\ \delta_2 &= a_2 x_1 + b_2\delta_1 \end{aligned}\right\}$$

and

$$\begin{vmatrix} c_2, & d_2 \\ a_2, & b_2 \end{vmatrix} = 1.$$

And there is a whole series of such relations, ending with expressions for x, δ, in terms of x_{n-1}, δ_{n-1}.

Hence, by the lemma of Article 15, it appears that there are relations, got by eliminating x_1, δ_1, ... x_{n-1}, δ_{n-1}, of the form

$$\left.\begin{aligned} x &= C'x_0 + D'\delta_0 \\ \delta &= A'x_0 + B'\delta_0 \end{aligned}\right\} \quad\text{......................(20)},$$

where the determinant of the coefficients C', D', A', B' equals the product of n determinants each of which is unity, and therefore is itself unity; in fact,

$$B'C' - A'D' = 1 \quad\text{......................(21)}.$$

These coefficients A', B', C', D' must of course depend on the values of u and v; they are not therefore constants depending merely on the nature of the instrument, but are constants for the instrument so long as we keep to the same first and last planes of reference. It is obvious that the relations between the entering and emerging rays must be quite independent of the particular method which we adopted of specifying the intermediate rays within the instrument, and therefore the constants do not depend on the particular choice of the intermediate planes of reference.

Of course y and ϵ are expressible in terms of y_0 and ϵ_0 by equations involving the same four constants.

17. Dependence of A', B', C', D' on u and v.

In order to interpret optically the equations (20), it is necessary to ascertain in what way the constants depend on u and v. With this in view we introduce a new set of constants, A, B, C, D, which are the values that A', B', C', D', respectively, would have if u and v were both zero. These are genuinely constants of the instrument, and correspond to planes of reference taken close up to the first and last refracting surfaces respectively. They are, of course, subject to the relation

$$BC - AD = 1 \quad \text{......................(22).}$$

If ξ_0, η_0 and ξ, η are the coordinates of the points where the ray meets the first and last surfaces of the instrument respectively, these coordinates must take the place of x_0, y_0, x, y in equations (20) when these surfaces and the corresponding planes of reference coincide. So we have the relations

$$\left.\begin{aligned} \delta &= A\xi_0 + B\delta_0 \\ \xi &= C\xi_0 + D\delta_0 \end{aligned}\right\} \quad \text{......................(23).}$$

The geometrical equations of the entering and emerging rays give relations between x_0 and ξ_0, and between ξ and x, namely

$$\left.\begin{aligned} \xi_0 &= x_0 + u\delta_0 \\ x &= \xi + v\delta \end{aligned}\right\} \quad \text{......................(24).}$$

The elimination of ξ_0 and ξ between (23) and (24) leads to

$$\left.\begin{aligned} \delta &= Ax_0 + (Au + B)\,\delta_0 \\ x &= (Av + C)\,x_0 + (Auv + Bv + Cu + D)\,\delta_0 \end{aligned}\right\} \quad \text{......(25).}$$

These must be identical with (20), and so we have the relations

$$\left.\begin{aligned} A' &= A \\ B' &= Au + B \\ C' &= Av + C \\ D' &= Auv + Bv + Cu + D \end{aligned}\right\} \quad \text{..................(26).}$$

The equations (25) are the most useful form of the equations of the instrument. The second alone involves the whole theory and may be called the fundamental equation. The first is really contained in the second, for it may be derived from it by differentiation with respect to v, it being seen from the geometrical equation that

$$\frac{\partial x}{\partial v} = \delta \quad \text{..........................(27).}$$

It appears therefore that the whole of the theory of the instrument, when aberration is neglected, depends on four "constants of the instrument," which are equivalent to only three independent constants on account of the relation (22).

The constant A is called the "power" of the instrument.

III. THE OPTICAL PROPERTIES OF A SYMMETRICAL INSTRUMENT.

18. Conjugate focal planes.

In proceeding to deduce from equations (25) some of the optical properties of the instrument, it is advantageous to think of the point (x_0, y_0) in the first plane of reference as the seat of a point source of light, and to regard the last plane of reference as occupied by a white screen.

The fundamental equation shews that the value of x in general depends on that of δ_0, so that rays setting out from the luminous point in different directions and passing through the instrument will strike the screen in different points. Thus there is a bright patch on the screen, but not an image. It is, however, possible, when the position of the bright point is prescribed, to choose such a value of v, and therefore such a position of the screen, that x shall be independent of δ_0; this is effected by choosing v so that the coefficient of δ_0 in the fundamental equation vanishes. The coefficient of ϵ_0 in the expression for y, being the same coefficient, vanishes for the same choice of v. And so the point in which the ray meets the screen is the same whatever be the direction in which it originally set out. Thus all the rays come together to form a point image at the point (x, y) in the final plane of reference.

The relation between the reduced distances u, v, of the bright point and its image, from the ends of the instrument is

$$Auv + Bv + Cu + D = 0 \qquad\qquad (28).$$

The value of v thus defined does not depend on x_0 or y_0, but only on u, and so a number of bright points having the same u give images having the same v. Consequently a small plane object placed on the axis in the plane defined by u has a plane image at the position defined by v. The relation between object and image is, in a sense, a reciprocal one, since light, proceeding from an object in the plane v and traversing the instrument in the reverse direction, would form an image in the plane u.

Planes which correspond to one another in this fashion are called "conjugate focal planes."

When A is not zero, the relation between conjugate focal planes may be put in another form if we multiply (28) through by A, and make use of (22). We get

$$\begin{aligned} 0 &= A^2uv + ABv + ACu + AD \\ &= A^2uv + ABv + ACu + BC - 1, \end{aligned}$$

whence

$$(Au + B)(Av + C) = 1 \qquad\qquad (29).$$

19. Conjugate foci.

A bright point and its image point are called "conjugate foci." The expression of the coordinates of the image point in terms of those of the object point to which it is conjugate is contained in the equations

$$x = \frac{x_0}{Au+B}, \quad y = \frac{y_0}{Au+B}, \quad v = -\frac{Cu+D}{Au+B} \quad \text{......(30)*},$$

which are immediate consequences of (25) and (28). Here u and v are coordinates, for origins in the first and second planes of reference respectively, in the special case in which μ_0 and μ are both unity; in the more general case u and v are z coordinates divided respectively by μ_0 and μ.

20. Linear Magnification.

The ratio of the linear dimensions of the image to those of the object is called the "linear magnification," and may be denoted by m. We may measure the object from the point on the axis to some other point (x_0, y_0); and then, when v has been adjusted so as to satisfy (28), the image is measured from that point of its plane which is on the axis, to the point (x, y). So far as regards dimensions parallel to the axis of x, the linear magnification is the ratio of x to x_0, and this is given by the second of equations (25). So we have

$$m = Av + C \quad \text{.........................(31).}$$

Combining this with (29) we get

$$\frac{1}{m} = Au + B \quad \text{.........................(32).}$$

These are the two magnification formulae; the symmetry of the instrument about the axis ensures that these formulae, proved in the first instance only for dimensions parallel to the axis of x, shall be true for dimensions in all directions perpendicular to the axis of the instrument.

A negative value of m indicates inversion of the image as compared with the object.

* The correspondence between a point (x', y', z'), referred to one set of axes, and a point (x, y, z), referred possibly to another set of axes, which is determined by the relations

$$\frac{x'}{a_1x+b_1y+c_1z+d_1} = \frac{y'}{a_2x+b_2y+c_2z+d_2} = \frac{z'}{a_3x+b_3y+c_3z+d_3} = \frac{1}{px+qy+rz+s},$$

is called a homographic correspondence. It is a one-to-one correspondence in which straight lines correspond to straight lines. Clearly the correspondence between conjugate foci is of this type.

Since m is the same for dimensions in all directions, the image is of the same shape as the object; for example the image of a circular object is circular. If we had precise image formation with an asymmetrical instrument, the kind referred to in Article 8, the linear magnification would be different for different directions and the image would be of a different shape from the object, i.e. distorted; a circular object, for example, would have an elliptic image.

21. Unit Planes or Principal Planes.

It is possible so to choose u and v that the linear magnification shall be unity. Denoting the special values by u_1 and v_1, we obtain, by putting $m=1$ in (31) and (32), the relations

$$1 = Av_1 + C, \quad 1 = Au_1 + B \quad \text{.................(33)}.$$

The conjugate focal planes defined by u_1 and v_1 are called the "planes of unit magnification," or more briefly "unit planes" or "principal planes." The points where these planes meet the axis are called the "unit points" of the instrument. The characteristic property of the unit planes may be expressed by the statement that the (x, y) coordinates of the point where any entering ray crosses the first unit plane are the same as those of the point where the corresponding emergent ray crosses the second unit plane.

22. Principal Focal Planes.

If in the relation (28) we put $u=\infty$, the corresponding value, V, of v is given by

$$AV + C = 0 \quad \text{.........................(34)};$$

if we put $v=\infty$, the corresponding value U of u is given by

$$AU + B = 0 \quad \text{.........................(35)}.$$

The planes specified by U and V are called the "principal focal planes" of the instrument, and their intersections with the axis are called the "principal foci." The second principal focal plane, that given by V, is the place where a screen should be placed in order to receive the image of an infinitely distant object; in other words it is the locus of the points of concurrence of emergent rays corresponding to incident pencils of parallel rays. The first principal focal plane, that given by U, is such that if a point source of light be placed at any point of it (near the axis), the rays from it will, after traversing the instrument, emerge as a parallel beam. If the direction of passage through the instrument were reversed, the rôles of these two planes would be interchanged.

23. Focal Lengths.

The distances of the principal focal planes beyond the corresponding unit planes are called the "focal lengths" of the instrument. The word "beyond" is used to indicate that distances are to be measured from the respective ends of the instrument, away from the instrument; i.e., opposite to the direction of the light for the end at which incidence takes place, and in the same direction as the light at the end where the light emerges.

The *reduced* distances of the principal focal planes beyond the corresponding unit planes are respectively

$$U - u_1 \text{ and } V - v_1,$$

each of which equals $-1/A$. The actual distances, or focal lengths, are therefore F_1, F_2, where

$$F_1 = -\mu_0/A \text{ and } F_2 = -\mu/A \quad \text{..............(36).}$$

These lengths are negative if the instrument has a positive power, and *vice versa*. They are equal if the first and last media have the same index; and in the particular case of $\mu_0 = 1$, $\mu = 1$, which is practically true for an instrument in air, the focal length is

$$F = -1/A \quad \text{..........................(37).}$$

24. Relation between the distances of conjugate foci from the principal focal planes.

When distances are measured from the principal focal planes instead of from the ends of the instrument the magnification formulae and the condition for conjugacy assume specially simple forms. Let $u - U = p$, $v - V = q$, so that p, q are the reduced distances of the planes of reference beyond the principal focal planes. Substituting $U + p$, $V + q$, for u, v, in formulae (26), and remembering (34) and (35), we obtain

$$A' = A, \quad B' = Ap, \quad C' = Aq, \quad D' = Apq - \frac{1}{A},$$

so that the standard formulae (20) assume the form

$$\left.\begin{aligned} \delta &= Ax_0 + Ap\delta_0 \\ x &= Aqx_0 + \left(Apq - \frac{1}{A}\right)\delta_0 \end{aligned}\right\} \quad \text{..................(38).}$$

From these it appears that the condition that x should be independent of δ_0, that is, the relation between conjugate focal planes, is

$$Apq - \frac{1}{A} = 0,$$

or

$$pq = (1/A)^2 \quad \text{..........................(39).}$$

The associated magnification formulae are

$$\frac{1}{m} = Ap, \quad m = Aq \quad \text{..........................(40).}$$

These formulae are, of course, equivalent to

$$\frac{1}{m} = -\frac{p'}{F_1}, \quad m = -\frac{q'}{F_2}, \quad p'q' = F_1F_2 \text{..............(41),}$$

where p', q' are the true distances corresponding to p, q. In this form they can be strikingly illustrated by a diagram in which the axes of the instrument, the unit planes, and the principal foci, B_1, B_2, are first drawn. From a point P of the object two rays are drawn, the first parallel to the axis and meeting the first unit plane in X, the second through B_1 meeting the first unit plane in Y. The corresponding emergent rays are a ray which leaves the second unit plane at X' and passes through B_2, and a ray which leaves the second unit plane at Y' and proceeds parallel to the axis; the point Q of intersection of these two marks the position of the image of P. The lines XX' and YY' are parallel to the axis, in virtue of the property of unit planes. All the lengths named in (41) are easily identified in the diagram, and the formulae are seen to be the expressions of very simple geometrical relations.

25. Nodal Points.

If an incident ray cuts the axis of the instrument at the point whose reduced distance from the first refracting surface is u, then for this ray $x_0 = 0$, and the first of equations (25) becomes

$$\delta = (Au + B)\,\delta_0.$$

Now if u have the value u_2 given by the relation

$$Au_2 + B = \mu/\mu_0 \text{..........................(42),}$$

we have

$$\delta/\mu = \delta_0/\mu_0,$$

so that the projected inclinations of the entering and the emergent rays are equal. The emergent ray must pass through the image point of the point on the axis determined by u_2, namely a point on the axis at reduced distance v_2 beyond the end of the instrument from which rays emerge, v_2 being given by

$$Av_2 + C = \mu_0/\mu \text{..........................(43).}$$

These two conjugate points on the axis are called the "nodal points" of the instrument. They have the property that if an entering ray passes through the first nodal point the emergent ray passes through the second nodal point and is parallel to the entering ray.

If $\mu = \mu_0$, it is seen by comparison of the formulae (33), (42), and (43) that $u_1 = u_2$ and $v_1 = v_2$. In other words, if the first and last media have the same index the nodal points are the points where the unit planes cut the axis.

Suppose the instrument to be directed towards a very distant bright point, and the image caught on a properly placed screen. If now the instrument be rotated about a point on its axis through a small angle, the image will still be (very approximately) on the screen, but not in general at the same point of the screen. One ray suffices to determine the position of the image, and we shall take that ray to be the ray through the nodal points. The object being very distant the direction of the entering ray is not altered by the rotation of the instrument, and so the direction of the ray that emerges and passes through the second nodal point is unaltered. Thus the image is shifted on the screen by an amount equal to the distance moved through by the second nodal point on account of the rotation of the instrument. If the point about which the rotation takes place is the second nodal point itself, that point does not move, and so the image does not move on the screen. This fact is taken advantage of in practice, to determine experimentally the nodal points of an optical instrument. The image is studied through a microscope while the instrument to be tested is turned successively about different points on the axis; when a point has been found such that a small rotation about it does not move the image it is known to be the nodal point.

26. Equivalent Single Refracting Surface.

The equations (26) shew how the constants, A', B', C', D', depend on the positions of the planes of reference. If we take the unit planes as planes of reference we must substitute the values of u_1 and v_1 for u and v in these equations, and the corresponding constants are

$$A' = A, \quad B' = 1, \quad C' = 1, \quad D' = 0 \quad \ldots\ldots\ldots\ldots(44);$$

so the equations (20) assume the form

$$\left.\begin{aligned} x &= x_0 \\ \delta &= Ax_0 + \delta_0 \end{aligned}\right\} \quad \ldots\ldots\ldots\ldots\ldots\ldots\ldots\ldots(45).$$

Let us compare these with the corresponding equations for a single refracting surface of power κ, when the planes of reference both coincide with the surface itself. The coincidence of the planes of reference ensures the equality $x = x_0$, and we have already seen (Article 11, eqn. 14), that

$$\delta = \kappa x_0 + \delta_0 \quad \ldots\ldots\ldots\ldots\ldots\ldots\ldots\ldots(46).$$

Thus it is seen that, if $\kappa = A$, the equations for the single surface are identical with those of the instrument; and, if the initial and final media have respectively the same indices in the two cases, the interpretations of corresponding symbols are identical. To a given entering ray, therefore, there correspond in the two cases emergent rays having the same direction and proceeding from points having the same (x, y) coordinates. The only difference is in the z coordinates of the points of departure from the second unit planes; these necessarily differ by the distance from the first to the second unit plane of the original instrument. So, for a given object at a given distance from the first unit plane, the single refracting surface of power A produces an image which differs only in position from the image produced by the instrument; the positions differ by a displacement parallel to the axis, of amount equal to the distance between the unit planes.

This single surface is called the "equivalent single refracting surface" of the instrument. When the object is in a prescribed position relative to the instrument, the equivalent surface must be regarded as situated at the first unit plane; but when the relative position of the object is not prescribed, so that a translation of object or of image presents no difficulty, one may think of the space between the unit planes as abolished and the equivalent surface as being in both.

27. Apparent distance of the image.

When the values of u and v, in the formulae (25), are not such as correspond to conjugate focal planes, the coefficient of δ_0 in the expression for x does not vanish. It has, however, a simple interpretation if we regard u as marking the position of the object, and v as corresponding to the point on the axis occupied by the eye of the observer.

An observer estimates the distance of an object at which he is looking, by the use of both eyes. If he is compelled to use only one eye, he loses the power of estimating distance from mere observation, but can usually combine the result of observation with a previous knowledge of the approximate size of the object, and so calculate its distance. What the one eye can perceive is the angular size of the cone of rays by which the object is seen, and what is calculated is the distance at which a normal section of this cone should be made if it is to be of the known linear dimensions of the object.

In viewing an object through a monocular instrument, there is usually no previous knowledge of the size of the image. It is natural therefore to attribute to the image the known size of the object, though its actual size may be very different. Thus there arises the idea of the "apparent distance" of the image, as distinguished from its true distance. The apparent distance of the image is the distance from the eye to that normal section of the cone of rays from image to eye whose linear dimensions are the same as those of the object.

Measuring the object from the axis of the instrument, its linear size is what has been called x_0. The angle that the corresponding image subtends at the eye, placed on the axis at v, is $-\delta/\mu$. There is, however, no necessity to take account of sign, since the apparent distance must be positive, i.e. a distance measured forwards from the eye. Change in the sign of δ simply means inversion of the image. The apparent distance is d, where

$$x_0/d = -\delta/\mu \quad \text{.........................(47).}$$

Since the eye is on the axis, $x = 0$, and hence, by formulae (25)

$$\delta = Ax_0 + (Au + B)\,\delta_0,$$

$$0 = (Av + C)\,x_0 + (Auv + Bv + Cu + D)\,\delta_0\,;$$

the elimination of δ_0 gives

$$x_0 = -(Auv + Bv + Cu + D)\,\delta,$$

whence

$$d = \mu\,(Auv + Bv + Cu + D) \quad \text{.................(48),}$$

the equality having reference to arithmetical rather than algebraical values.

When $\mu = \mu_0$, i.e. when object and eye are in the same medium, the apparent distance is not altered if the positions of object and eye be interchanged.

28. Longitudinal Magnification.

The linear magnification m has reference only to the dimensions of object and image in directions perpendicular to the axes of the instrument. The longitudinal magnification, which we may denote by M, is the ratio of the dimensions of image and object in the direction of the axis, the object being no longer regarded as flat.

Let u, u' be the least and greatest reduced distances of points of the object from the instrument, and v, v' the corresponding reduced distances for the image; and let m, m' be the corresponding values of the linear magnification. Clearly the longitudinal dimensions of object and image are $\mu_0(u' - u)$ and $\mu(v' - v)$, so that $M = \mu(v' - v)/\mu_0(u' - u)$.

Now

$$Av' + C = \frac{1}{Au' + B}, \quad Av + C = \frac{1}{Au + B},$$

whence $$v' - v = -(u'-u)/(Au'+B)(Au+B);$$
therefore $$M = -\mu/[\mu_0(Au'+B)(Au+B)] = -(\mu/\mu_0)\,mm' \quad\text{......(49)}.$$
When the longitudinal dimensions of the object are small we may assume $m' = m$ as a good approximation, and so we get
$$M = -(\mu/\mu_0)\,m^2 \quad\text{......................(50)}.$$

Hence, if the object is a curved surface cutting the axis normally, so also is the image; and the radii of curvature ρ_0, ρ, of their sections by any axial plane are in the same sense and are connected by the relation $\mu_0\rho_0 = \mu\rho$.

29. Helmholtz's formula.

The Helmholtz magnification formula expresses the linear magnification in terms of the inclinations of the ray which passes through the points where object and image meet the axis. If δ_0, δ refer to this ray, we have, by putting $x_0 = 0$ in the first of the standard formulae,
$$\delta = (Au+B)\,\delta_0,$$
or
$$m\delta = \delta_0 \quad\text{.............................(51)}.$$
This is the form assumed by Helmholtz's formula when expressed in terms of the notations of the present tract, to the degree of approximation postulated at the outset.

30. Other formulae.

The reduced distances u, v of the fundamental formulae, instead of being measured from the first and last surfaces of the instrument, may be measured from any other fixed planes of reference. Let us measure them from the pair of conjugate planes defined by u_0, v_0, for which the linear magnification is m_0, so that $m_0^{-1} = Au_0 + B$, and $m_0 = Av_0 + C$. We effect the change by substituting $u_0 + u$ and $v_0 + v$ for u and v respectively in (26). We thus get
$$A' = A,\quad B' = Au + m_0^{-1},\quad C' = Av + m_0,\quad D' = Auv + m_0^{-1}v + m_0 u.$$
Hence the relation between conjugate focal planes, $D' = 0$, takes the form
$$\frac{1}{m_0 u} + \frac{m_0}{v} + A = 0 \quad\text{..........................(52)},$$
while the magnification formulae become
$$m^{-1} = Au + m_0^{-1},\quad m = Av + m_0 \quad\text{..............(53)},$$
from which follows Maxwell's formula (equivalent to formula 49),
$$v/u = -mm_0 \quad\text{.........................(54)}.$$

This group of formulae includes those usually given for direct refraction at a single spherical surface. If u, v are measured from the surface itself, $m_0 = 1$. If u, v are measured from the centre of the sphere (the point of coincidence of the nodal points), $m_0 = \mu_0/\mu$.

IV. THE MANNER IN WHICH THE OPTICAL PROPERTIES OF AN INSTRUMENT DEPEND ON ITS CONSTITUTION.

31. Dependence of the values of the constants on the composition of the instrument.

With a view to determining the manner in which the constants of an instrument depend upon its structure, let us examine the effect of modifying the instrument already considered by adding to it another refracting surface. At the place marked by $v = a'$ let us place a refracting surface of power κ', beyond which shall lie a new medium of index μ', and let us evaluate the constants A_1, B_1, C_1, D_1, of the modified instrument.

Let x', y' be the coordinates of the point where the ray crosses the surface κ', and let δ', ϵ', be the reduced projected inclinations of the emergent ray. In the equations (25) let us put $u = 0$, $v = 0$, so that x_0 marks the point where the ray crosses the first refracting surface, and x the point where it crosses the last refracting surface of the unmodified instrument. Thus

$$\left.\begin{aligned} \delta &= Ax_0 + B\delta_0 \\ x &= Cx_0 + D\delta_0 \end{aligned}\right\} \quad \text{.........................(55).}$$

In terms of these, x' is expressed by the geometrical equation

$$x' = x + a'\delta,$$

and δ' by the optical equation

$$\delta' = \delta + \kappa' x'.$$

Eliminating x and δ, we get

$$\left.\begin{aligned} \delta' &= \{A + \kappa'(Aa' + C)\}\, x_0 + \{B + \kappa'(Ba' + D)\}\, \delta_0 \\ x' &= (Aa' + C)\, x_0 + (Ba' + D)\, \delta_0 \end{aligned}\right\} \quad \text{...(56).}$$

These equations correspond, in the case of the modified instrument, to the equations (55) for the unmodified instrument, and hence the constants for the modified instrument are given by

$$\left.\begin{aligned} A_1 &= A + \kappa'(Aa' + C) \\ B_1 &= B + \kappa'(Ba' + D) \\ C_1 &= Aa' + C \\ D_1 &= Ba' + D \end{aligned}\right\} \quad \text{..................(57).}$$

These forms shew explicitly in what manner the constants depend on the power κ' of the last refracting surface, and on the reduced distance

a' of the last surface from the last but one. The first of these relations as it stands, and the derived relations

$$C_1 = \frac{\partial A_1}{\partial \kappa'}, \quad D_1 = \frac{\partial B_1}{\partial \kappa'} \quad \text{.......................(58)},$$

are important.

The power A of an instrument, regarded not as a constant but rather as a function of the powers, $\kappa_1, \kappa_2, \ldots \kappa_n$, and the mutual reduced distances, $a_1, a_2, \ldots a_{n-1}$, of the component refracting surfaces, may be called K, with a suffix to indicate the number of surfaces in the instrument. Then in the case of an instrument of n surfaces the relations (58) become

$$A = K_n, \quad C = \frac{\partial K_n}{\partial \kappa_n}, \quad D = \frac{\partial B}{\partial \kappa_n} \quad \text{..............(59)}.$$

Now suppose the ray to go through the instrument in the reverse direction, the axis of z being likewise reversed. The change leaves x unaltered, but converts a positive reduced projected inclination into a negative one. Solving equations (55), and remembering that the determinant of the constants is unity, we get

$$\left. \begin{aligned} -\delta_0 &= Ax + C(-\delta) \\ x_0 &= Bx + D(-\delta) \end{aligned} \right\} \quad \text{....................(60)},$$

where now $(-\delta)$ and $(-\delta_0)$ are reduced projected inclinations of the reversed ray. These equations shew that for the reversed rays the A and D constants play the same part as before, but the rôles of B and C are interchanged. The last refracting surface is now κ_1, and so we have further equations of the same type as (59), namely

$$B = \frac{\partial K_n}{\partial \kappa_1}, \quad D = \frac{\partial C}{\partial \kappa_1} \quad \text{.......................(61)},$$

of which the latter is equivalent to

$$D = \frac{\partial^2 K_n}{\partial \kappa_1 \partial \kappa_n} \quad \text{.........................(62)}.$$

Thus when the functional form of K_n or A is known, the forms of the other constants can be immediately deduced.

The first of equations (57) gives us a method of finding the function K_n. For, in the new notation, this equation becomes

$$K_n = K_{n-1} + \kappa_n \left(K_{n-1} a_{n-1} + \frac{\partial K_{n-1}}{\partial \kappa_{n-1}} \right) \quad \text{...........(63)},$$

a formula for K_n in terms of K_{n-1}.

From formulae (14) it is clear that, for a single surface of power κ_1,

$$K_1 = \kappa_1 \quad \text{............................(64)}.$$

Now put $n=2$ in (63), and we find that for an instrument consisting of two surfaces $(\kappa_1, a_1, \kappa_2)$

$$K_2 = \kappa_1 + \kappa_2(\kappa_1 a_1 + 1) = \kappa_1 + \kappa_2 + a_1\kappa_1\kappa_2 \quad \text{.............(65)}.$$

Similarly

$$\left.\begin{aligned} K_3 &= (1+\kappa_3 a_2)(\kappa_1+\kappa_2+a_1\kappa_1\kappa_2) + \kappa_3(1+a_1\kappa_1) \\ &= \kappa_1+\kappa_2+\kappa_3+a_1\kappa_1\kappa_2+(a_1+a_2)\kappa_1\kappa_3+a_2\kappa_2\kappa_3+a_1a_2\kappa_1\kappa_2\kappa_3 \end{aligned}\right\}\text{...(66)},$$

and the process may be continued to any stage by repeated application of the general formula.

Another rule for writing down K_n may be got from the equations (57). For it is clear from these equations that

$$\left.\begin{aligned} A_n &= \kappa_n C_n + A_{n-1} \\ B_n &= \kappa_n D_n + B_{n-1} \\ C_n &= a_{n-1} A_{n-1} + C_{n-1} \\ D_n &= a_{n-1} B_{n-1} + D_{n-1} \end{aligned}\right\} \quad \text{.................(67)};$$

on comparison of these with the formulae of simple continued fractions it is readily seen that in the case of the fraction

$$\frac{1}{\kappa_1+}\,\frac{1}{a_1+}\,\frac{1}{\kappa_2+}\,\frac{1}{a_2+}\,\frac{1}{\kappa_3}+\dots \quad \text{..............(68)},$$

the convergents corresponding to the partial quotients a_{n-1} and κ_n are respectively D_n/C_n and B_n/A_n. This is a convenient form in which to remember the law of formation of K_n; but the student must not suppose that familiarity with the theory of continued fractions is at all necessary to the understanding of the theory of the optical instrument.

32. Thick and thin lenses.

The simplest instrument made up of two refracting surfaces is the lens. If κ_1, κ_2, be the powers of the surfaces, and a the reduced distance between them, the constants of the instrument are

$$A = \kappa_1+\kappa_2+a\kappa_1\kappa_2, \quad B = 1 + a\kappa_2, \quad C = 1 + a\kappa_1, \quad D = a \text{...(69)}.$$

Thus $\quad u_1 = -a\kappa_2/(\kappa_1+\kappa_2+a\kappa_1\kappa_2), \quad v_1 = -a\kappa_1/(\kappa_1+\kappa_2+a\kappa_1\kappa_2),$

and the distance between the unit planes is

$$\mu_0 u_1 + \mu_1 a + \mu v_1$$

or $$a\left[(\mu_1-\mu)\kappa_1 + (\mu_1-\mu_0)\kappa_2 + \mu_1 a\kappa_1\kappa_2\right]/\left[\kappa_1+\kappa_2+a\kappa_1\kappa_2\right] \text{......(70)}.$$

For a lens of glass, of thickness t, situated in air, we put $\mu_0 = \mu = 1$. If the radii of curvature of the faces be ρ_1 and ρ_2, positive when the faces are concave, $\kappa_1 = (\mu_1 - 1)/\rho_1$, $\kappa_2 = (\mu_1 - 1)/\rho_2$, and $a = t/\mu_1$. If t is so small that the squares of t/ρ_1 and t/ρ_2 may be neglected, the distance between the unit planes is sufficiently well represented by the expression $t\left(1-\frac{1}{\mu_1}\right)$, which is $\frac{1}{3}t$ if we assume for glass the rough value $\mu_1=\frac{3}{2}$. The power is accurately

$$\frac{\mu_1-1}{\rho_1\rho_2}\left\{\rho_1+\rho_2+t\left(1-\frac{1}{\mu_1}\right)\right\} \quad\text{..................(71).}$$

In ordinary simple cases the positions of the unit planes are easily specified. Thus if the lens is plano-concave one unit point is in the curved surface, and the other distant from it by a third of the thickness. If $\rho_1=\rho_2$, the unit points are approximately the points of trisection of the thickness.

A thin lens is a lens for which t is so small in comparison with ρ_1 and ρ_2 that $a\kappa_1$ and $a\kappa_2$ may be neglected. For such a lens $A=\kappa_1+\kappa_2$, and u_1, u_2, and the distance between the unit planes all vanish*. The unit planes being coincident, the lens may be replaced by the equivalent single refracting surface, without any shifting of either object or image, and the equivalence is not qualified but complete. Of course, if the first and last media are the same, air for example, a single refracting surface between air and air is not a practical possibility; but what is meant by the equivalent single surface in this case is that rays passing through the thin lens undergo precisely the same deviations as are indicated by the mathematical formulae for a single surface of power $\kappa_1+\kappa_2$ in the position occupied by the lens. We have in fact

$$\delta'-\delta=(\kappa_1+\kappa_2)\,x,$$

which becomes identical with (14) if $\kappa=\kappa_1+\kappa_2$.

33. Instrument made up of thin lenses.

An instrument made up of n thin lenses is, of course, an instrument of $2n$ refracting surfaces, and is a case of the theory already discussed. But we have seen that the pair of refracting surfaces which make up any thin lens can be replaced by a completely equivalent single refracting surface. Thus the mathematical treatment which applies most simply to this case is one which takes account

* Another expression for the power of a thin lens may be noted in passing, namely

$$A=(\mu_1-1)\left(\frac{1}{\rho_1}+\frac{1}{\rho_2}\right)=2(\mu_1-1)\frac{t'-t}{r^2},$$

where t' is the thickness at the rim of the lens, and r the radius of the circular rim.

of only n refracting surfaces, whose powers and mutual distances are the powers and mutual distances of the corresponding lenses. In fact the whole discussion of Articles 16—32 may be reinterpreted as applying to an instrument made up of n thin lenses, each κ being now the power of a lens instead of the power of a single surface. The expression for K_n as a function of the powers of the lenses is the same as we have already obtained in the case of single surfaces; and we may talk of the "equivalent single thin lens" of the instrument instead of the "equivalent single refracting surface"; the former being in fact a physical possibility in the particular case in which the latter is not, namely the case when $\mu = \mu_0$. If, as usually happens, all the lenses are situated in air, for which $\mu = 1$, the distinction between true and reduced distances disappears.

In particular it may be noticed that the mathematical formulae for a thick lens have precisely the same form as the formulae for an instrument made up of two thin lenses.

34. Instrument of zero power.

An instrument for which the constant A is zero is called a "telescope." Clearly its principal focal planes, unit planes, and nodal points are all infinitely distant, and the instrument has properties quite distinct from those of an instrument whose power is not zero.

To obtain these properties, we put $A = 0$ in the standard formulae (22) and (25), the result being

$$\left.\begin{aligned} BC &= 1 \\ \delta &= B\delta_0 \\ x &= Cx_0 + (Bv + Cu + D)\,\delta_0 \end{aligned}\right\} \quad \ldots\ldots\ldots\ldots\ldots(72).$$

From the last equation it is seen that x is independent of δ_0 provided

$$Bv + Cu + D = 0 \quad \ldots\ldots\ldots\ldots\ldots\ldots\ldots\ldots(73),$$

which is therefore the condition of conjugacy of the planes determined by u and v. This relation is simpler than the general relation

$$Auv + Bv + Cu + D = 0,$$

for while the general relation is the most general homographic relation between u and v, the particular relation is a special kind of homographic relation, namely similitude. The general relation implies equality of anharmonic ratios in the case of four u's and the corresponding v's; the particular relation implies equality of ordinary ratios of division in the case of three u's and the corresponding v's.

Assuming u and v to correspond to conjugate focal planes, we get from the third of equations (72) the magnification formula

$$m = \frac{x}{x_0} = C \quad \text{..........................(74)},$$

which, in virtue of the first equation, is equivalent to

$$\frac{1}{m} = B \text{..................................(75)}.$$

Thus it appears that the linear magnification is the same, at whatever distance along the axis the object be placed.

From equation (73) it is seen that if $u = \infty$, then also $v = \infty$. The principal focal planes are thus both at infinity. In other words, to an entering pencil of parallel rays there corresponds an emergent pencil of parallel rays. The relation between the directions of these pencils is determined by the second of equations (72). Suppose two objects, whose distances are (for practical purposes) infinite, to be in the field of view of the instrument, for example two stars, one on the axis and the other near the axis. If θ_0 is the angle which these two stars would subtend at the eye of an observer, i.e. the angular distance between them, θ_0 is the angle between the beams of light from them which enter the instrument; thus $\theta_0 = \delta_0/\mu_0$. The corresponding beams of light which emerge from the instrument and enter the eye are inclined to one another at an angle θ, equal to δ/μ, and this θ is the angular distance between the images of the stars. The ratio of θ to θ_0 is the angular magnification (n) produced by the instrument; it is given by the formula

$$n = \frac{\theta}{\theta_0} = \frac{\mu_0}{\mu} \cdot \frac{\delta}{\delta_0} = \frac{\mu_0}{\mu} B \text{........................(76)}.$$

If, as is usually the case, $\mu = \mu_0$, we get the still simpler formula

$$n = B = \frac{1}{m} \quad \text{..........................(77)}.$$

Thus the angular magnification is the inverse of the linear magnification, and is independent of the angular positions of the objects viewed, provided only they be near enough to the direction of the axis of the telescope to justify the use of the approximation which we have adopted throughout.

An ordinary astronomical telescope in ideal adjustment consists of an object-glass and an eye-piece placed at such a distance apart that the power of the combined instrument is zero. The linear magnification produced by the telescope being independent of the position of the object, we may think of an object consisting of a bright disc of the same size as the object-glass, placed close up to the latter. The image of this formed by the object-glass only (regarded as a thin lens) is an equal bright disc, close up to the object-glass on the inside. The final image, formed by the eye-piece, is thus the same as the image of the object-glass formed by the eye-piece—i.e. what is called the eye-ring. So the linear magnification is the ratio of the diameter of the eye-ring to the diameter of the object-glass, and the angular magnifying power is the inverse of this ratio.

Another formula for the magnifying power is got by applying formula (40) to the eye-piece only, so that the constants are the constants of the eye-piece. If f is the focal length of the eye-piece, and F that of the object-glass, we put $A=-1/f$; and we notice that in ideal adjustment eye-piece and object-glass have a common principal focal plane, so that for the object-glass or any object coincident with it $p=F$. The formula then yields

$$\frac{1}{m}=-\frac{F}{f}.$$

Here m, a particular linear magnification for the eye-piece alone, is the general linear magnification for the whole instrument. So the angular magnifying power is

$$n=\frac{1}{m}=-\frac{F}{f}\qquad\qquad(78).$$

35. Plane refracting surface.

A plane refracting surface has zero power, and therefore has the properties of a telescope. It is exceptional in this respect that, since its linear magnifying power is unity, its unit planes are not necessarily at infinity—they are indeterminate; any pair of conjugate planes are unit planes.

It is worth while to notice that if one of the refracting surfaces of an instrument be plane, it need not be reckoned as one of the n surfaces in the general formulae, provided care be taken to estimate properly the reduced distance, through it, between the refracting surfaces adjacent to it on either side. Thus, if we consider an instrument made up of two hemispheres of glass (radius r, index μ), with their plane faces towards each other at a distance c apart, we have really an instrument of four refracting surfaces. But we may treat it as one having only two refracting surfaces (κ_1, a_1, κ_2) provided we put $\kappa_1=\kappa_2=(1-\mu)/r$, $a_1=r/\mu+c+r/\mu$. If the curved faces of the hemispheres were towards each other at a distance c apart, we should put $\kappa_1 = \kappa_2 = (1-\mu)/r$, $a_1 = c$, and measure u, v, as reduced distances out through the glass from the curved surfaces; for example if the object were at a distance p from the first plane refracting surface, we should have $u = p + r/\mu$; and if the object were a virtual object (i.e. a real image which the entering rays would have formed if they had not been intercepted by the glass) p would be negative, but the same formula for u would be employed.

36. Field-glasses and telescopes.

Telescopes, field-glasses and opera-glasses are so constructed that the distance between the objective and the eye-piece can be adjusted to suit the eye of the observer. Probably these instruments, when in use, are seldom ideally telescopic, i.e. of zero power. It might, however, be thought that, since the adjustment to suit the average eye is a fairly small relative displacement of the lenses, the instruments in use are at least approximately telescopic, or of very large focal length. But this assumption requires examination, the more so as it is meaningless to call a length very large unless one can specify some standard length in comparison with which the largeness is estimated.

The linear magnification of an instrument in air, for an object at a distance u, is given by

$$1/m = Au + B,$$

and it is to be noticed that the right-hand side of this equality consists of two terms, one constant, the other involving u. If the instrument were telescopic the term containing u would be absent. And therefore, as regards linear magnification, it may be said that the instrument is *nearly* telescopic if the term in $1/m$ which involves u is small compared with the constant term, i.e. if Au/B is small so that the linear magnification is approximately independent of u. If, for example, the instrument be a Galileo's telescope, the focal lengths of objective and eye-glass being F and $-f$, and the distance between the lenses $F-f-b$,

$$A = b/Ff,\ B = (F-b)/f,\ \text{and}\ Au/B = bu/F(F-b) = bu/F^2$$

approximately. We should call the instrument nearly telescopic if bu/F^2 were small, or (in other words) if the distance between the principal foci of the lenses were small compared with the distance of the image formed by the object-glass from the second principal focus of the object-glass.

Measurements, rather rough because no proper optical bench was available, were made by the author on a pair of field-glasses of the ordinary type. The values obtained were $F = 7\cdot3$ inches, $f = 1\cdot4$ inches, and when the instrument was at the greatest length permitted by the manner in which it was mounted the distance between the lenses was 5·9 inches, so that b was zero and the adjustment ideally telescopic. Distant objects could be seen clearly with the instrument so adjusted, but the clearness was increased by bringing the lenses nearer together and was best when the instrument had been shortened by about ·2 inch. With this value of b, $bu/F^2 = u/266$, when u is measured by inches. As one does not use field-glasses to view an object whose distance is small compared with 22 ft., the instrument cannot be regarded as nearly telescopic, as regards its linear magnification, when put to its ordinary use.

The linear magnification, however, is not so important as the angular magnification, i.e. the ratio of the angles subtended at the eye by image and object. When the length of the instrument is denoted by l, and the distance of the eye from the eye-glass is negligible, the angular magnification is the ratio of $-x/v$ to $x_0/(l+u)$, that is $-m\,(l+u)/v$; it is easily verified that this is equal to

$$\{1 + lu^{-1}\}/\{C + Du^{-1}\} \quad \text{.....................(79).}$$

For the two-lens instrument which we have been discussing, this formula becomes

$$(1 + lu^{-1})/\{(f+b)\,F^{-1} + lu^{-1}\},$$

and if we assume that the length of the instrument is small compared with u, we see that approximately

$$\text{angular magnification} = F/(f+b) \quad \text{...........(80).}$$

In ideal adjustment this takes the value F/f; thus the change from ideal adjustment alters the angular magnification in the ratio $f/(f+b)$. If $f = 1\cdot4$ in., $b = \cdot2$ in., as in the case mentioned above, this ratio is 7/8; thus the change of angular magnification is not great, and the instrument is in this respect not very different from an ideal telescope.

In practice the value of b depends partly on the distance of the object and partly on the particular distance between eye and image to which the eye of the observer most readily accommodates itself. The value of b is $f^2(-v-f)^{-1} - F^2(u-F)^{-1}$. If the eye prefers a near image, so that $-v$, though greater than f, is not very great, b is positive and A is positive. But if the eye prefers a very distant image, $-v$ may be much greater than u, so that b and A are negative. The mounting of the lenses usually permits both positive and negative values of b.

37. Combination of two instruments.

When two instruments whose constants are known are placed end to end, so that light traverses first one and then the other, the constants of the resulting compound instrument can be calculated easily if each of the original instruments be supposed replaced by its equivalent single surface (or single thin lens). In the resultant two-surface instrument the distance between the single surfaces must be taken equal to the distance between the second unit plane of the first instrument and the first unit plane of the second. The object must, however, be regarded as having the same position relative to the single surface replacing the first instrument as it actually has relatively to the first unit plane of the first instrument.

A knowledge of the constants of an instrument is equivalent to knowledge of its power and of the positions of its unit planes. Taking the unit planes of an instrument of power κ as planes of reference, the constants for those planes are κ, 1, 1, 0. If we have two instruments of powers κ_1 and κ_2 whose adjacent unit planes are at reduced distance a apart, we see, by substituting the equivalent single surfaces, that the constants of the combined instrument are $\kappa_1 + \kappa_2 + a\kappa_1\kappa_2$, $1 + a\kappa_2$, $1 + a\kappa_1$, a, the planes of reference to which these constants refer being the first unit plane of the first instrument, and the second unit plane of the second instrument, since these planes correspond to the first and last refracting surfaces of the equivalent two-surface system.

It is worth noting that the constants for a combination of two instruments, which we may suppose to be situated in air, assume simple forms if expressed in terms of the focal lengths f_1, f_2 of the instruments, and the distance b between their adjacent principal foci. The planes of reference being the first principal focal plane of the first instrument, and the second principal focal plane of the second instrument, we readily obtain:

$$A = A' = \frac{b}{f_1 f_2}, \quad B' = -\frac{f_1}{f_2}, \quad C' = -\frac{f_2}{f_1}, \quad D' = 0 \quad \ldots(81).$$

For a combination of three instruments the corresponding results are

$$A = \frac{f_2^2 - b_1 b_2}{f_1 f_2 f_3}, \quad B' = \frac{b_2 f_1}{f_2 f_3}, \quad C' = \frac{b_1 f_3}{f_1 f_2}, \quad D' = -\frac{f_1 f_3}{f_2} \ldots(82).$$

V. THE EQUIVALENT THIN LENS.

38. Advantages of a compound instrument.

In connexion with the fact that any symmetrical optical instrument in air may, so far as the first order approximation is concerned, be replaced by an equivalent single thin lens, there naturally presents itself the question—what advantage has the relatively complex instrument over the equivalent single lens?

The most obvious answer to this question is that, by adjustment of the curvatures of the faces of the lenses that make up the instrument, it is possible to diminish greatly the aberrations of rays that are inclined to the axis, and so to increase the angular field which is represented after refraction by a clear image. Aberration is an effect depending on the small quantities of the second or higher orders which have been neglected throughout the present discussion. The effect of diminishing aberration is to extend the values of the small inclinations for which the formulae we have obtained may be regarded as approximately representing the actual facts.

But, if we agree to pay no attention to second order small quantities, we may still ask whether the single thin lens is really as effective as the instrument it replaces. The first telescopes were not corrected for aberration, and there must be some more elementary reason why they were constructed of two lenses when a single lens might have sufficed.

For a telescope in ideal adjustment there is, of course, no equivalent thin lens. For the power is zero, and the unit planes are at infinity. To give such relative translation to the object and image systems as shall bring these unit planes into coincidence, and then to use a lens of zero power, is a proceeding which can hardly be regarded as practicable. But the Galileo telescope, or opera-glass, as frequently used, is not a telescope in ideal adjustment, and might therefore be expected to have an equivalent thin lens capable of being used in its stead.

The magnification formula $1/m = Ap$, where p is the distance of the object beyond the first principal focal plane, shews that, when an instrument is used for viewing objects beyond the first principal focus, the linear magnification cannot be positive and so the image cannot be upright unless the power of the instrument is positive. So, when the opera-glass is used in this way, the power is positive and the equivalent thin lens is a diverging one. Now if we view an object

AB through a diverging lens, and if $A'B'$ be the image, and O the centre or unit-point of the lens, $OA'A$ and $OB'B$ are straight lines, and the image (if upright) is on the same side of the lens as the object, namely on the side remote from the eye; if the object is distant, the image is nearer to the eye than the object.

The eye E is necessarily outside the triangle AOB, and consequently the image $A'B'$ subtends at the eye a smaller angle than

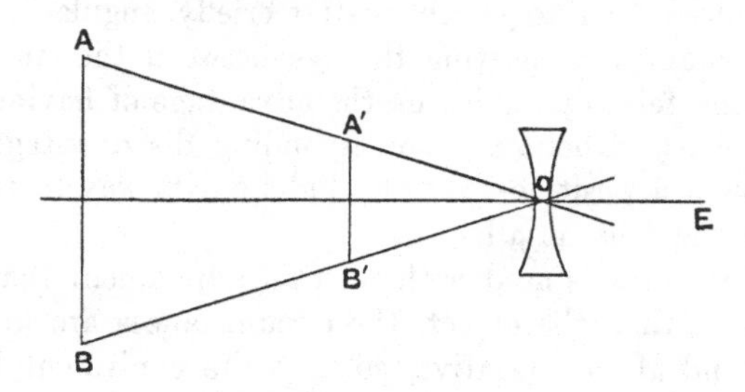

the object AB; thus there is no angular magnification, but rather angular diminution, and the lens is not serving any useful end. If only E could be on the same side of the lens as AB, the angle $A'EB'$ would be greater than the angle AEB, and angular magnification would be attained; but it is impossible to use a lens if the eye is on the same side of it as the object.

Now let us consider a Galileo's telescope in which the object-glass is converging and of focal length F, the eye-glass diverging and of focal length f, and the distance between the lenses $F-f-b$. Using the standard formulae for two thin lenses, we put $\kappa_1 = -1/F$, $\kappa_2 = 1/f$, $a = F-f-b$; whence

$$\left.\begin{array}{l} A = \kappa_1 + \kappa_2 + a\kappa_1\kappa_2 = b/Ff, \\ B = 1 + a\kappa_2 = (F-b)/f,\ C = 1 + a\kappa_1 = (f+b)/F \end{array}\right\} \quad \ldots\ldots(83).$$

In order that the power of the instrument may be positive, b must be positive, i.e. the lenses are nearer together than in the ideal adjustment. The distance of the second unit plane from the eye-end of the instrument is $(1-C)/A$ or $f(F-f-b)/b$, which is positive, and usually large compared with f. Now the eye is placed quite close to the eye-glass, and therefore the unit plane is *behind* the eye. The eye is in front of the unit plane for the eye-end of the instrument, and yet the instrument is between the eye and the object, so that it can be used for viewing the object. Now the eye and the image system occupy the same position relatively to the equivalent thin lens as if that lens were in the second unit plane, and so the use made of the instrument corresponds to the use of the single lens when the eye is between the lens and the object, under which circumstances there is a genuine angular magnification. The difference between the two cases is that for such a position of the eye the rays *do not* traverse the single lens before reaching the eye, but *do* traverse the telescope; the single lens is as useless as if it were absent, the telescope gives the image magnified as required. Or, to put the matter briefly, angular magnification can only be obtained by putting the eye between the single lens and the object; the telescope gives us the advantage of having the eye in front of the lens, without the corresponding disadvantage of having the lens in such a position that the light on its way to the eye does not traverse the lens at all.

If the opera-glass is used with such an adjustment that the image is more distant than the object, the circumstances are different. In this case b and A are negative, so that the equivalent lens is converging and would yield angular magnification to an eye placed behind it. The single lens would, in fact, be used like a reading-glass or a simple microscope. If d, D denote the distances of lens and object from the eye, and ϕ the focal length, $\phi + d > D$, and the angular magnification n is given by $n^{-1} = 1 - d(D-d)\phi^{-1}D^{-1}$. Thus a value of n sensibly greater than unity requires both d and ϕ to be of the same order of greatness as D. For distant objects this is clearly impracticable. But with a compound instrument it is easy to arrange for the equivalent conditions, namely that the power be small and the eye at a great distance behind the second unit plane.

39. The Unilens.

There has lately been invented an instrument which seems at first to constitute a refutation of what we have said about the practical uselessness of the equivalent thin lens. Major Baden Powell's "Unilens" is a single lens which, when held at a distance of three or four feet from the eye, performs the functions of a field-glass and gives an upright magnified image of distant objects. The action of the unilens is very interesting; it is a single convex lens of about six feet focal length, so that when it is being used the eye is between the lens and the principal focus. The image of a distant object is therefore a real inverted image, behind the principal focus and therefore behind the eye. The pencils which are on their way to form this image are intercepted by the lens of the eye, and the eye accommodates itself to receive the resulting image on the retina. The accommodation required is of the same kind as that which prepares the eye to view infinitely distant objects, but it is exaggerated to an extent which may be called accommodation for objects beyond infinity. The manufacturers state that only three people out of four can use the unilens; the use of the instrument is found to tire the eye.

VI. REFLECTING INSTRUMENTS.

40. Analytical formulae expressing the laws of reflexion.

The laws of reflexion are :—(i) the incident ray, the reflected ray, and the normal to the reflecting surface at the point of incidence are in one plane; (ii) the angle of incidence is equal to the angle of reflexion.

Using the notation of Article 1, with this modification that (l', m', n') shall be the cosines of the *reversed reflected* ray, we see that the diagram of that Article would be suitable for a reflexion if OC were equal to OB and inclined to OK at the same angle as OB, but on the other side of OK; in fact if C were on BJ produced, and $CJ = JB$. Now this amounts simply to a law of refraction in which $\mu' = -\mu$, and so we get the laws of reflexion by putting this value of μ' in equations (1). The result is

$$\left.\begin{aligned} l' + l &= 2L\cos\phi \\ m' + m &= 2M\cos\phi \\ n' + n &= 2N\cos\phi \end{aligned}\right\} \quad\ldots\ldots\ldots\ldots\ldots\ldots(84).$$

When all the inclinations to the axis of z are very small we have the approximate formulae

$$\left.\begin{aligned} l' + l &= 2L \\ m' + m &= 2M \end{aligned}\right\}\ldots\ldots\ldots\ldots\ldots\ldots\ldots\ldots(85),$$

and when the mirror is a spherical surface whose radius ρ is reckoned positive when the convexity is towards the positive direction of the axis of z,

$$L = x/\rho, \quad M = y/\rho,$$

and
$$l' + l = 2x/\rho, \quad m' + m = 2y/\rho \;\ldots\ldots\ldots\ldots\ldots(86),$$

where (x, y) are coordinates of the point of incidence.

Formulae (86) are useful and easy to remember, but for our present purpose it is convenient to modify them so that their form may correspond with that of the formulae for a refracting surface. Since l' is the projected inclination of the reversed reflected ray to the axis of z, it is clear that, if we take an axis of z' in the opposite direction to the axis of z, i.e. in the direction of the reflected central ray, the corresponding projected inclination of the actual reflected ray is $-l'$. This, multiplied by the index μ of the medium in which the ray travels, is the reduced projected inclination (to the axis of z') of the reflected ray. We may put $-2\mu/\rho=\kappa$, and call κ the power of the mirror, noting that κ is positive when the reflecting surface is convex to the rays, and that κ depends on the index of the medium in which the mirror is placed.

With this notation the typical formula becomes

$$\delta'=\delta+\kappa x \quad \text{......................(87)},$$

which differs from the refraction formula only in the interpretation of δ'.

41. Single Spherical Mirror.

In studying image formation by a single mirror, we take planes of reference for the incident and refracted rays at reduced distances u, v, respectively from the mirror. The coordinates of the points where the rays cut these planes are (x_0, y_0) and (x, y). Then by mathematical reasoning formally identical with that for a single refracting surface (Article 14), we get

$$\left.\begin{aligned}\delta'&=\kappa x+(\kappa u+1)\,\delta\\ x'&=(\kappa v+1)\,x+(\kappa uv+u+v)\,\delta\end{aligned}\right\} \quad \text{..............(88)}.$$

Thus the condition for conjugate focal planes is

$$\kappa uv+u+v=0 \quad \text{......................(89)},$$

and the magnification formulae are

$$m^{-1}=\kappa u+1, \quad m=\kappa v+1 \quad \text{....................(90)}.$$

42. Geometrical interpretation.

There is a remarkable difference between the geometrical significance of formula (89) when applied to a mirror and when applied to a refracting surface. This is made clear if we replace reduced distances by true distances. In the case of the mirror the index of the medium is really irrelevant, and if we denote true distances by accented letters we find that the condition for conjugate focal planes is

$$-2\rho^{-1}u'v'+u'+v'=0 \quad \text{...................(91)}.$$

On the other hand for a refracting surface we get

$$\kappa u'v'+\mu' u'+\mu v'=0 \quad \text{......................(92)}.$$

The latter formula represents homography, the former represents a very special kind of homography, namely involution.

That the homography represented by (91) is an involution is a fact which, if it were not mathematically obvious from the form of the equation, would be inferred at once from the physically obvious reciprocal relation between an object and its image formed by reflexion. It might be thought that a thin lens must also yield an involution relation between conjugate foci for a similar reason. But though, in a sense, it is true that object and image are interchangeable for any refracting instrument, this interchange always implies a corresponding reversal of the direction in which the rays pass; if we insist on the rays always passing in the same sense, the relation between object and image is not a reciprocal one. From the purely mathematical point of view it might be argued that in the case of a thin lens, $\mu = \mu'$, and the coefficients of u' and v' in (92) are therefore equal; that is true, but as u' and v' are measured in opposite senses from the lens their coefficients would have to be equal in magnitude but opposite in sign in order that the homography should be an involution. It is otherwise with formula (91), since u' and v' are measured in the same sense from the mirror.

The double points of the involution of (91) are clearly given by

$$u' = v' = 0 \text{ and } u' = v' = \rho,$$

that is the apex of the mirror and the centre of curvature. For one double point $m = +1$, for the other $m = -1$.

43. Reflecting Telescopes.

As an example of the application of the reflexion formulae, let us consider the instrument known as Cassegrain's telescope. The incident light first strikes a concave mirror, is reflected backwards to strike a convex mirror, and is again reflected; the light after the second reflexion passes through a small hole in the middle of the concave mirror and is refracted through a lens which constitutes the eye-piece. If the radius of the concave mirror be R, that of the convex mirror r, the power of the lens κ, the distance between the mirrors a, and the distance from the convex mirror to the lens b, we have the same mathematical formulae as for a refracting instrument of three lenses, the values of the powers and mutual distances being

$$\kappa_1 = -2/R, \quad \kappa_2 = 2/r, \quad \kappa_3 = \kappa, \quad a_1 = a, \quad a_2 = b.$$

The power of the whole instrument is accordingly, by (66),

$$A = (1 + \kappa b)\left(-\frac{2}{R} + \frac{2}{r} - \frac{4a}{Rr}\right) + \kappa\left(1 - \frac{2a}{R}\right),$$

which must vanish if the instrument is to be in ideal telescopic adjustment. In that particular case the angular magnification would be

$$B = (1 + a_1\kappa_2)(1 + a_2\kappa_3) + a_1\kappa_3$$
$$= (1 + 2ar^{-1})(1 + \kappa b) + a\kappa.$$

In Gregory's telescope the second mirror is concave. The formulae are derived from the formulae for Cassegrain's telescope by changing the sign of r.

If the eye-piece used in either case is compound, i.e. consists of more than one lens, the formulae still apply if κ is the power of the eye-piece as a whole, and b the distance from the second mirror to the first unit plane of the eye-piece.

There are somewhat neater expressions for the constants in terms of the focal lengths of the mirrors and the eye-piece, and the distances between adjacent principal focal planes. They may be got by substitution in the formulae quoted at the end of Article 37.

44. Refracting instrument and single mirror.

From the point of view of pure geometry the formation of images by light which first traverses a refracting instrument, is then reflected by a mirror, and then goes back through the refracting instrument, is of some interest. The power of the instrument being A, and that of the mirror A', and the reduced distance of the mirror from the second unit plane of the instrument being a, it is easy to see by considering the equivalent single surface, that, save for a reversal of general direction, the emergent light behaves as if it had traversed three surfaces whose powers and distances, in order, are A, a, A', a, A. And therefore if u, v be measured, both in the same direction, from the first unit plane of the refracting instrument, the condition for conjugate focal planes is

$$Kuv + \frac{\partial K}{\partial \kappa_1} v + \frac{\partial K}{\partial \kappa_3} u + \frac{\partial^2 K}{\partial \kappa_1 \partial \kappa_3} = 0,$$

where K is the function K_3 formed from

$$\kappa_1 = A, \quad a_1 = a, \quad \kappa_2 = A', \quad a_2 = a, \quad \kappa_3 = A.$$

By the symmetry of K, the coefficients of u and v in this relation are equal; hence it represents an involution. On physical grounds this was to be expected, since the relation between object and image is a reciprocal one. The unit planes of the compound instrument clearly coincide, namely at one of the double points of the involution; the other double point corresponds to magnification equal to -1, for an interchange of object and image changes m to $1/m$, and therefore when object and image coincide it is necessary that $m = 1/m$, or $m^2 = 1$. The instrument can be replaced by a single equivalent mirror at the unit plane, with its centre at the other double point of the involution; and the equivalence is mathematically complete without any translation of object or image, since the unit planes coincide.

VII. ENTRANCE AND EXIT PUPILS.

45. Amplitude of the pencil of rays from any point.

The pencil of rays, proceeding from any bright point, which traverses the whole of an optical instrument, is liable to be restricted in consequence of the necessarily restricted area of each of the refracting or reflecting surfaces of which the instrument is composed. Thus the area of a thin lens available for transmission of light is determined by the circular edge of the lens, or by the aperture of the opaque frame in which the lens is held ; the pencil cannot be of greater width than that necessary to fill this aperture. Some one of these opaque boundaries (not necessarily that of smallest aperture) will restrict the transmitted pencil to a greater extent than would any of the others, and will thus be effective in determining the angular extent of the emergent pencil. Or it may be that somewhere on the axis of the instrument there is placed an opaque "stop" with a circular hole in it, through which the rays must pass, and that this hole is so narrow that its amplitude is what determines the extreme rays of the transmitted pencil.

The effective "stop," whether it be coincident with one of the refracting or reflecting surfaces, or not, has its aperture completely filled with rays. The image of this aperture (treated as a bright object) formed by the whole of the series of surfaces at which the rays suffer reflexion or refraction after traversing the stop, is a circular area perpendicular to the axis of the instrument which is completely filled by the emergent rays (or by these rays produced backwards if necessary). This circular area is called the "exit-pupil." The extreme rays of the pencil converging to any image point are got by joining the image point to all the points of the boundary of the exit-pupil.

The image of the effective stop which would be got by reversing the general direction of the rays and using only the surfaces which lie between the stop and the object-end of the instrument, is called the "entrance-pupil." The extreme rays of that part of the pencil, which enters the instrument from any bright point, which is destined ultimately to emerge, are got by joining the bright point to all the points of the boundary of the entrance-pupil.

Since the formation of a good image depends on keeping the inclinations of the rays very small, there is an obvious advantage in the use of a stop of small aperture. But there are the corresponding disadvantages of a more restricted field of view and a diminution in the quantity of the transmitted light.

VIII. CHROMATIC DEFECTS OF THE IMAGE.

46. Dispersion.

Light is a periodic wave motion of the aether, but ordinary white light is not a simply periodic motion having a single period, like the motion of a simple pendulum; it resembles, rather, the more complicated oscillations of a dynamical system having several degrees of freedom, such as a number of particles attached at different points to a string which hangs from a fixed end; the motion is very complex, but may be regarded as compounded of a number of distinct motions each of which is simply periodic, which have different periods. Aethereal oscillations of different periodicities produce in the eye different colour sensations, and white light is not light of any particular pure colour, but is a mixture of lights of different periodic times or different colours, namely the colours of the rainbow and the spectrum.

In free aether and in air lights of different colours are propagated with the same velocity, but in other transparent media the light-disturbances corresponding to different colours are propagated with different velocities and have different indices of refraction. Consequently any given refracting optical instrument is likely to produce different effects upon beams of light of different colours; and to an entering beam of mixed (e.g. white) light there will correspond, not a single emergent beam of mixed light, but a number of emergent beams of different colours probably in different directions and having different focal properties. This is most strikingly exemplified in Newton's method of producing a spectrum of sunlight by means of a glass prism. The phenomenon of the separation of the beams of different colours is called "dispersion."

Instruments composed entirely of reflecting surfaces are, of course, free from dispersion effects.

47. Dispersive power of a substance.

In estimating the extent to which a substance behaves differently towards light-beams of different colours, we fix our attention on two standard colours in different parts of the spectrum, for example blue and red, and we measure the refractive index for each of these colours. The difference between the two indices, or any simple function proportional to this difference, might be taken as a measure of the dispersive power so far as the selected colours are concerned. Of course the standard colours have to be more precisely defined than by mere colour-names; they are identified by their positions in the spectrum, being the colours corresponding to selected bright lines in the spectra of familiar substances. For our purpose it will suffice to call them "colour No. 1" and "colour No. 2," and to denote the corresponding indices by μ_1 and μ_2.

The deviation produced by a prism of small angle a on rays which are incident nearly normally is known to be approximately $(\mu - 1)\,a$. Hence the difference between the deviations of rays of the two standard colours is $(\mu_2 - \mu_1)\,a$. The ratio of the difference of deviations to the absolute deviation of a ray of a third standard colour (No. 3) is therefore

$$(\mu_2 - \mu_1) / (\mu_3 - 1),$$

and depends, not on the shape, but only on the nature of the material of the prism. It is therefore a suitable quantity to take as a measure of the extent to which the material behaves differently to the two standard colours, and is called the "dispersive power" of the substance. It is usually written

$$(\triangle\mu)/(\mu - 1) \quad\text{.........................}(93),$$

the operator symbol $\triangle$ being used to denote the increment of the function of μ on which it operates when μ is changed from μ_1 to μ_2, and the suffix being dropped in the case of colour No. 3. For brevity the dispersive power is denoted by the symbol ϖ. It must be noted that a numerical value of ϖ for any particular substance is meaningless unless it is accompanied by a precise definition of the standard colours to which it refers.

The dispersive power ϖ is usually a very small quantity. When colour No. 1 corresponds to the C line (in the red), colour No. 2 to the F line (in the blue), and colour No. 3 to the D line (in the yellow), the values of ϖ for some different kinds of Jena glass are roughly 1/67, 1/55, 1/65, 1/51, 1/44.

48. Powers of a thin lens for different colours.

The power of a thin lens of glass, situated in air, is given by the formula

$$\kappa = (\mu - 1)\left(\frac{1}{\rho} - \frac{1}{\sigma}\right),$$

where ρ and σ are the radii of curvature of the faces. Hence, for two standard colours,

$$\kappa_2 - \kappa_1 = (\mu_2 - \mu_1)\left(\frac{1}{\rho} - \frac{1}{\sigma}\right).$$

Dividing corresponding sides of these equalities we get

$$\frac{\kappa_2 - \kappa_1}{\kappa} = \frac{\mu_2 - \mu_1}{\mu - 1} = \varpi,$$

or

$$\triangle \kappa = \kappa\varpi \quad \text{.............................(94).}$$

This formula gives the difference between the powers, for the standard colours, of a lens made of a material whose dispersive power is known.

49. Achromatism.

Since, in general, a refracting optical instrument behaves differently to light-beams of different colours, it follows that it has not simply one set of constants *A*, *B*, *C*, *D*, but many such sets, a set for each colour. And if objects which differ only in colour be placed successively in the same position in front of the instrument, the corresponding images will usually be in different positions and of different sizes. If the object be white, or if the light proceeding from it be a mixture of several pure colours, the instrument forms not one image, but a series of images, in different positions, of different sizes and different colours. The eye perceives all these images simultaneously, the resulting appearance being usually a somewhat blurred image of the same mixed colour as the object, surrounded by a party-coloured fringe.

This defect of the image is called "chromatic aberration," and an instrument which is free or nearly free from the defect is said to be "achromatic."

In ideal achromatism the instrument would be such that the images of all colours would be of the same size and in the same position. But ideal achromatism is difficult of attainment, and an instrument-maker is usually satisfied if he can cause the images of two standard colours to coincide. When this has been effected it is generally the case that the images of all the colours which lie between the standard colours in the spectrum are much more nearly coincident than they would be in the absence of the two-colour achromatism, and the same is true to some extent for the other parts of the spectrum. Thus the image formed by an instrument which is to this extent achromatic is very much better than would be formed by an instrument designed without regard to achromatism.

The most valuable device for the partial removal of chromatic defects is the combination, in one instrument, of different kinds of glass, having different dispersive powers. If the dependence of index on colour were such as to give the same value to the ratio

$$(\mu_p - \mu_q)/(\mu_p - \mu_r)$$

for different kinds of glass, for every selection of three colours (p, q, r), then any instrument composed of these kinds of glass, if achromatic for two colours, would be achromatic for all colours. The fact that the law of dispersion has not actually this simple form is referred to as "the irrationality of dispersion."

The particular pair of colours for which the instrument is made achromatic must be chosen with reference to the use to which the instrument is to be put. If the instrument is to be used for vision, the colours selected are those which, partly on account of the relatively large extent to which they are present in white light and partly on account of the greater sensitiveness to them of the nerves of the eye, are most important in ordinary vision; such colours are those in the neighbourhood of Fraunhofer's C and F lines. But if the instrument is to be used only for photographic purposes, the chemical activity of the colours becomes an important consideration, while their effect on the eye no longer matters; the colours selected would therefore be in the violet or ultra-violet part of the spectrum.

50. Partial achromatism.

Even for two colours, achromatism is not easy to obtain, and sometimes the instrument-maker is content with partial achromatism. Complete achromatism requires the two coloured images to be of the same size and in the same position; if one of these conditions is satisfied but not the other, the achromatism is partial. The eye is a better judge of size than of distance, and so if only one of the conditions can be satisfied it is better to achromatize for size than for position.

It is further to be observed that an instrument may be achromatic for a particular position of the object without being achromatic for all positions of the object. This "particular" achromatism is of course easier to obtain than the "general" achromatism which holds for all positions of the object.

51. Achromatism of a symmetrical instrument composed of thin lenses in air.

So far as concerns the image formation corresponding to the first-order approximation assumed throughout the present Tract, the conditions of achromatism can be deduced immediately from the four-constant formulae. Taking the magnification formulae,

$$\frac{1}{m} = Au + B, \quad m = Av + C,$$

we remark first that for an instrument in air the first and last media have their indices unity for all colours. Hence u, v are the true distances of object and image from the instrument. The object being in the same position for all colours, $\Delta u = 0$. Achromatism for size requires that $\Delta m = 0$ and therefore of course $\Delta \frac{1}{m} = 0$. Achromatism for position of the image requires $\Delta v = 0$. If therefore we apply the operator Δ to the magnification formulae we see that

$$u \Delta A + \Delta B = 0 \quad \text{.........................(95)}$$

is the condition for achromatism as regards size, and

$$v \Delta A + \Delta C = 0 \quad \text{.........................(96)}$$

a further condition if the achromatism is for both size and position.

If the achromatism is only for a particular position of the object, i.e. for a particular value of u, these equations as they stand represent the conditions. But if the achromatism is to be for all values of u, we clearly require

$$\Delta A = 0, \quad \Delta B = 0 \quad \text{.......................(97)}$$

for size, with the further condition

$$\Delta C = 0 \quad \text{...........................(98)}$$

for position*.

It is a well known theorem that if $f(x_1, x_2, x_3 \ldots)$ is a function of the independent variables $x_1, x_2, x_3, \ldots$, having continuous partial differential coefficients, and if the symbol Δ be used for *small* increments,

$$\Delta f = \frac{\partial f}{\partial x_1} \Delta x_1 + \frac{\partial f}{\partial x_2} \Delta x_2 + \frac{\partial f}{\partial x_3} \Delta x_3 + \ldots + \omega,$$

where ω is small of a higher order than the increments; hence if ω be omitted from the right-hand side we have a statement of equality which is very approximately true.

* Since $BC - AD = 1$, the vanishing of ΔA, ΔB, and ΔC involves also the vanishing of ΔD.

Now the constants of the instrument under consideration are functions of the powers of the component thin lenses and their mutual true distances, and these distances are not affected by change from one colour to another. So the operator $\triangle$ affects only the κ's, and $\triangle\kappa$ is in each case small. Accordingly we can apply the theorem of the total differential, just quoted, and we have

$$\begin{aligned}\triangle A &= \frac{\partial A}{\partial \kappa_1}\triangle\kappa_1 + \frac{\partial A}{\partial \kappa_2}\triangle\kappa_2 + \ldots + \frac{\partial A}{\partial \kappa_n}\triangle\kappa_n,\\ &= \frac{\partial A}{\partial \kappa_1}\kappa_1\varpi_1 + \frac{\partial A}{\partial \kappa_2}\kappa_2\varpi_2 + \ldots + \frac{\partial A}{\partial \kappa_n}\kappa_n\varpi_n \text{ (by Art. 48)} \ldots(99),\end{aligned}$$

with corresponding expressions for $\triangle B$ and $\triangle C$. Substituting these in the formulae (95) and (96), or (97) and (98), we have the conditions for partial or complete, particular or general, achromatism.

52. Instrument consisting of two lenses.

For an instrument consisting of two thin lenses, κ_1, κ_2, at a distance a apart,

$$A = \kappa_1 + \kappa_2 + a\kappa_1\kappa_2, \quad B = 1 + a\kappa_2, \quad C = 1 + a\kappa_1,$$

$$\left.\begin{aligned}\triangle A &= (1 + a\kappa_2)\,\kappa_1\varpi_1 + (1 + a\kappa_1)\,\kappa_2\varpi_2\\ \triangle B &= a\kappa_2\varpi_2, \quad \triangle C = a\kappa_1\varpi_1\end{aligned}\right\} \ldots\ldots\ldots(100).$$

The conditions for general achromatism as to the size of the image, namely $\triangle A = 0$, $\triangle B = 0$, are obviously equivalent to $\varpi_1 = 0$, $\varpi_2 = 0$. Accordingly this partial achromatism for all positions of the object is unattainable unless each of the thin lenses separately is achromatic, i.e. made of a substance which has no dispersive power. But the only known substances whose dispersive power is practically zero are those whose index is not sensibly different from unity, and lenses made of such material would have no power. So general achromatism as to size is unattainable with two thin lenses at a distance apart.

The condition for achromatism as to size, for a particular position of the object, is

$$\{(1 + a\kappa_2)\,\kappa_1\varpi_1 + (1 + a\kappa_1)\,\kappa_2\varpi_2\}\,u + a\kappa_2\varpi_2 = 0 \ \ldots\ldots(101),$$

which assumes a still simpler form if the lenses are of the same material, so that $\varpi_1 = \varpi_2$. It is always possible to choose a, κ_1, κ_2 so that this shall be satisfied for a prescribed value of u.

53. Achromatism of the eye-piece of a telescope.

If the objective of a telescope be a reflector, or a compound object-glass which is itself achromatic, it is necessary that the eye-piece also should be an achromatic instrument. When the telescope is in ideal adjustment the achromatism required in the eye-piece is of a particular kind, corresponding to a particular position of the object. For the object is the real image formed by the objective, and is in the first principal focal plane of the eye-piece.

For this position of the object it is meaningless to talk of achromatism for size or achromatism for position, inasmuch as the final image is infinitely distant and the linear magnification is infinite.

The pencil of rays proceeding from a particular point of the object in the focal plane becomes, after refraction through the eye-piece, a beam of parallel rays whose direction determines the angular position of the image perceived by the eye. If the beams of different colours enter the eye in different directions, there will be a series of images of different colours and in different angular positions, in fact a chromatic effect which would lessen the utility of the telescope. The kind of achromatism to be aimed at is therefore parallelism of the emerging beams of different colours.

This statement is however only approximately correct, for the eye-piece has different focal planes for different colours. The object is accordingly in the focal plane for one colour, but out of the focal plane for all other colours. Only one of the emerging beams is a beam of parallel rays, the others consist of rays which are in each case very nearly parallel. The condition of achromatism is therefore more precisely stated as parallelism of the principal or central rays of the emerging pencils of different colours.

The formula for the direction of an emerging ray is

$$\delta = Ax_0 + (Au + B)\,\delta_0,$$

the notation being as usual. In the present application u has such a value as to place the object in the first focal plane of the eye-piece, x_0 refers to a particular point of the object, and δ_0 to a particular ray proceeding from that point.

We shall apply the operator $\triangle$ to both sides of this equation, noting that though for one colour $Au + B$ is zero it is not necessarily zero for any other colour. As x_0 and δ_0 are unaffected by $\triangle$, the result of the operation is

$$\triangle\delta = x_0 \triangle A + (u \triangle A + \triangle B)\,\delta_0\,;$$

the condition of achromatism is that, when δ_0 corresponds to the central ray of the pencil,

$$\triangle\delta = 0.$$

Thus the condition of achromatism is

$$x_0 \triangle A + (u \triangle A + \triangle B)\, \delta_0 = 0,$$

when a suitable value has been substituted for δ_0.

It is usual, though not correct, to take as the central ray of the pencil the ray which, before entering the eye-piece, is parallel to the axis. For this ray $\delta_0 = 0$, and therefore the corresponding condition for achromatism is

$$\triangle A = 0 \qquad \text{(102).}$$

Thus we have the condition usually given for the achromatism of the eye-piece of a telescope in ideal adjustment, namely that the powers of the eye-piece for the standard colours shall be the same.

In the case of a pencil of nearly parallel rays, it cannot matter much which of the rays is taken as the principal or central ray, and therefore the condition just obtained must give very good achromatism. But it is undoubtedly more precise to treat the ray through the centre of the objective of the telescope as the principal ray. For this ray $\delta_0 = x_0/F$, where F is the focal length of the objective, and so the condition for achromatism is

$$\triangle A + (u \triangle A + \triangle B)\, \frac{1}{F} = 0,$$

or if we put $u + F = b$, so that b is the distance between the objective and the field-lens of the eye-piece,

$$\triangle A + \frac{1}{b} \triangle B = 0 \qquad \text{(103).}$$

Let us consider, as an example, an eye-piece consisting of two thin lenses κ_1, κ_2, at a distance a apart. For this eye-piece

$$A = \kappa_1 + \kappa_2 + a\kappa_1\kappa_2,$$

and the usual condition of achromatism is

$$0 = \triangle A = (1 + a\kappa_2)\, \kappa_1 \varpi_1 + (1 + a\kappa_1)\, \kappa_2 \varpi_2 \qquad \text{(104).}$$

On the other hand $B = 1 + a\kappa_2$, $\triangle B = a\kappa_2\varpi_2$, and the more precise condition for achromatism is

$$(1 + a\kappa_2)\, \kappa_1 \varpi_1 + \left(1 + a\kappa_1 + \frac{a}{b}\right) \kappa_2 \varpi_2 = 0 \qquad \text{(105).}$$

The formulae are the same if we agree to neglect a/b, the ratio of the distance between the lenses of the eye-piece to the distance between eye-piece and objective. This is usually quite a small quantity.

The usual condition of achromatism assumes a simple form if the lenses of the eye-piece are of the same kind of glass. When this is the case $\varpi_1 = \varpi_2$, and the condition reduces to

$$\kappa_1 + \kappa_2 + 2a\kappa_1\kappa_2 = 0.$$

Put $\kappa_1 = -1/f_1$, $\kappa_2 = -1/f_2$, so that f_1, f_2 are the focal lengths of the lenses, positive when converging, and we get the still simpler result

$$2a = f_1 + f_2 \quad \text{.........................(106)},$$

or, in words, the distance between the lenses is the arithmetic mean of their focal lengths. This condition is satisfied by the Huygens eye-piece.

54. Achromatism of object-glasses.

For an instrument consisting of two thin lenses close together the constants are

$$A = \kappa_1 + \kappa_2, \quad B = 1, \quad C = 1.$$

Clearly there is only one necessary condition for two-colour achromatism, namely

$$\triangle A = 0 \quad \text{.........................(107)}.$$

When this is satisfied the system is achromatic for size and position of the image, wherever the object may be.

The condition is equivalent to

$$\kappa_1\varpi_1 + \kappa_2\varpi_2 = 0 \quad \text{.........................(108)}.$$

In order to construct an instrument of this kind, of given power A, and achromatic for the two standard colours, it is only necessary that the powers κ_1, κ_2 of the component lenses should be the quantities determined by the equations

$$\kappa_1 + \kappa_2 = A, \quad \varpi_1\kappa_1 + \varpi_2\kappa_2 = 0 \quad \text{.................(109)}.$$

It is to be remarked, however, that the two lenses must not be of the same kind of glass, for if $\varpi_1 = \varpi_2$ the second equation becomes $\kappa_1 + \kappa_2 = 0$, which is inconsistent with the first. The possibility of constructing an achromatic object-glass of power different from zero depends on the availability of two kinds of glass having different dispersive powers.

For an instrument consisting of three thin lenses close together, the constants are $A = \kappa_1 + \kappa_2 + \kappa_3$, $B = 1$, $C = 1$, and the sole condition of achromatism

$$\kappa_1\varpi_1 + \kappa_2\varpi_2 + \kappa_3\varpi_3 = 0 \quad \text{....................(110)}.$$

If, therefore, we consider three standard colours, and use ϖ for the dispersive power of a substance with reference to colours No. 1 and No. 2, and ϖ' for the dispersive power with reference to colours No. 1

and No. 3, we can construct an object-glass achromatic for the three standard colours and of given power A, by choosing κ_1, κ_2, κ_3 so as to satisfy the equations

$$\left.\begin{aligned}\kappa_1+\kappa_2+\kappa_3&=A\\ \varpi_1\kappa_1+\varpi_2\kappa_2+\varpi_3\kappa_3&=0\\ \varpi_1'\kappa_1+\varpi_2'\kappa_2+\varpi_3'\kappa_3&=0\end{aligned}\right\}\quad\ldots\ldots\ldots\ldots\ldots\ldots(111).$$

But the possibility of doing this depends on there being available kinds of glass having such values of ϖ, ϖ', as shall allow of the consistence of these equations. The objective of the Sheepshanks instrument in the Cambridge Observatory is achromatic for three colours. It may perhaps seem unfair to apply to the components of the objective of a great telescope the theory of thin lenses, but it is to be remembered that all that is here meant by the term "thin" is that the thickness is small in comparison with the radii of curvature of the surfaces.

55. Instrument not composed of thin lenses.

For the sake of generality we consider a symmetrical refracting instrument in which not only the intermediate media but also the media outside the instrument at either end are possessed of dispersive power.

The magnification formulae

$$\frac{1}{m}=Au+B,\quad m=Av+C,$$

are now better rewritten in the form

$$\frac{1}{m}=\left(\frac{A}{\mu_0}\right)u'+B,\quad m=\left(\frac{A}{\mu}\right)v'+C,$$

where u', v' are true, not reduced, distances. The operator Δ leaves u' unaltered; and achromatism for size requires $\Delta m=0$, $\Delta\left(\frac{1}{m}\right)=0$; achromatism for position as well as size requires the further condition $\Delta v'=0$.

Hence it appears that the condition of achromatism for the size of the image is

$$u'\Delta\left(\frac{A}{\mu_0}\right)+\Delta B=0\quad\ldots\ldots\ldots\ldots\ldots\ldots\ldots(112),$$

and that the further condition for achromatism as to the position of the image is

$$v'\Delta\left(\frac{A}{\mu}\right)+\Delta C=0\quad\ldots\ldots\ldots\ldots\ldots\ldots\ldots(113).$$

The expansions of such expressions as $\triangle\left(\frac{A}{\mu_0}\right)$ are much more complicated than in the case of thin lenses, for each power κ_r depends on two indices, μ_{r-1} and μ_r, and each reduced distance a_r is of the form a'_r/μ_r, where a'_r is the corresponding true distance. Thus

$$\triangle a_r = -\frac{a'_r}{\mu_r^2}\triangle\mu_r = -\frac{\mu_r - 1}{\mu_r} a_r \varpi_r \quad \text{...........(114)};$$

and, as

$$\kappa_r = \frac{\mu_r - \mu_{r-1}}{\rho_r},$$

$$\triangle\kappa_r = \frac{\triangle\mu_r - \triangle\mu_{r-1}}{\rho_r}$$

$$= \frac{(\mu_r - 1)\varpi_r - (\mu_{r-1} - 1)\varpi_{r-1}}{\mu_r - \mu_{r-1}}\kappa_r \quad \text{...........(115)}.$$

Consequently for any function $f(\kappa_1, a_1, \ldots \kappa_n)$,

$$\triangle f = -\sum_r \frac{\mu_r - 1}{\mu_r} a_r \varpi_r \frac{\partial f}{\partial a_r} + \sum_r \frac{(\mu_r - 1)\varpi_r - (\mu_{r-1} - 1)\varpi_{r-1}}{\mu_r - \mu_{r-1}}\kappa_r \frac{\partial f}{\partial \kappa_r} \quad \text{...(116)}.$$

It appears, therefore, that the general conditions of achromatism involve heavy formulae. In the case of simple instruments, such as the thick lens, it is best to discard general formulae and proceed from first principles.

56. Single thick lens in air.

Let the radii of curvature of the surfaces of a lens of index μ be ρ_1 and ρ_2, both positive if the surfaces are convex; and let the thickness be t. We shall examine what condition must be satisfied if the lens is to have the kind of achromatism required in the eye-piece of a telescope.

The power is

$$A = -\frac{\mu - 1}{\rho_1} - \frac{\mu - 1}{\rho_2} + \frac{t}{\mu}\cdot\frac{(\mu - 1)^2}{\rho_1\rho_2},$$

$$= \frac{\mu - 1}{\rho_1\rho_2}\left\{\frac{\mu - 1}{\mu}t - (\rho_1 + \rho_2)\right\} \quad \text{.................(117)};$$

whence, when squares of $\triangle\mu$ are neglected as usual,

$$\frac{\triangle A}{A} = \frac{\triangle\mu}{\mu - 1} - \frac{\triangle\mu}{\mu} + \frac{(t - \rho_1 - \rho_2)\triangle\mu}{(\mu - 1)t - \mu(\rho_1 + \rho_2)}.$$

The condition for the particular kind of achromatism contemplated is $\triangle A = 0$, which is equivalent to

$$\varpi \left[\frac{1}{\mu} + \frac{(t - \rho_1 - \rho_2)(\mu - 1)}{(\mu - 1)t - \mu(\rho_1 + \rho_2)} \right] = 0,$$

or

$$t = \frac{\mu^2}{\mu^2 - 1} (\rho_1 + \rho_2) \quad \text{.....................(118).}$$

If the lens has this particular thickness it will be achromatic.

The procedure by which this result has been obtained is that which usually leads most quickly to the desired condition. In the present instance it happens to be easier to equate A_1 to A_2 without using the $\triangle$ operator. We thus get the more precise form of the condition, namely

$$t = \frac{\mu_1 \mu_2}{\mu_1 \mu_2 - 1} (\rho_1 + \rho_2) \quad \text{.....................(119).}$$

57. Jena glasses*.

With a view to the construction of achromatic combinations of lenses, a large number of kinds of glass having different optical properties are manufactured at Jena, and numbers specifying these properties are given in the catalogue published by the manufacturers.

For this purpose the indices of refraction are measured for five different colours which are identified by means of bright lines in familiar spectra; these are a red potassium line $K\alpha$ denoted by A', a red hydrogen line $H\alpha$ denoted by C, a yellow sodium line $Na\alpha$ denoted by D, and two other hydrogen lines, $H\beta$ which is greenish blue and $H\gamma$ which is blue, denoted by F and G' respectively. The lines are enumerated in descending order of wave length; the interval from A' to G' is not far short of an octave.

The dispersive properties of a particular glass, so far as these colours are concerned, can be specified by giving the absolute value of the index for the line which lies near to the brightest part of the visible spectrum, namely D, and the differences of the indices for the four intervals CF, $A'D$, DF, FG'. As the interval CF contains the part of the spectrum which affects the eye most powerfully, the dispersive power is taken to be that corresponding to these two lines, namely $\varpi = (\mu_F - \mu_C)/(\mu_D - 1)$. This number is small, but if we put $1/\varpi = \nu$ we get a fairly large number, which is chosen in preference to ϖ for tabulation in the catalogue.

* Most of the numbers quoted in this and the following Article are taken from Müller-Pouillet's *Lehrbuch der Physik*, Bd. II.

The following table exhibits the optical constants of a few of the Jena glasses, the numbers at the heads of the columns being those which distinguish the respective glasses in the catalogue.

Kind of Glass	*O152*	*S8*	*S30*	*O60*	*O164*	*O103*
μ_D	1·5368	1·5736	1·5760	1·5179	1·5503	1·6202
$\mu_F - \mu_C$	·01049	·01129	·00884	·00860	·01114	·01709
ν	51·2	50·8	65·2	60·2	49·4	36·2
$\mu_D - \mu_{A'}$	·00659	·00728	·00570	·00553	·00710	·01034
$\mu_F - \mu_D$	·00743	·00795	·00622	·00605	·00786	·01220
$\mu_{G'} - \mu_F$	·00610	·00644	·00500	·00487	·00644	·01041

These figures illustrate some of the statements which have formed the basis of our discussion of dispersion. They shew, for example, that ν, and therefore also the dispersive power ϖ, is different for different kinds of glass, a fact which is at the very foundation of the theory of achromatism. If it were not so, it would be impossible to remove dispersion by a combination of prisms or lenses without at the same time doing away with the deviation or refractive effect which it is desired to retain; the formulae for an achromatic objective, $\kappa_1 + \kappa_2 = A$, $\varpi_1\kappa_1 + \varpi_2\kappa_2 = 0$, would be inconsistent unless A were zero, in which case the objective would be useless.

The table illustrates also the irrationality of dispersion; if we compare the first two columns we find that the mutual ratios of $\mu_F - \mu_C$, $\mu_D - \mu_{A'}$, etc. are slightly different for the glasses *O*152 and *S*8. But it is not so for every pair of glasses; in the case of *S*8 and *S*30 it is easily verified that there is almost perfect proportionality between the corresponding differences of index. In a combination of these two glasses, achromatism for two colours would carry with it achromatism for all the five standard colours.

In order to shew more clearly to what extent this desirable proportionality of the differences of corresponding indices obtains, the catalogue tabulates three other optical constants, namely

$$\alpha = \frac{\mu_D - \mu_{A'}}{\mu_F - \mu_C}, \quad \beta = \frac{\mu_F - \mu_D}{\mu_F - \mu_C}, \quad \gamma = \frac{\mu_{G'} - \mu_F}{\mu_F - \mu_C};$$

the more nearly these constants are equal in the case of two kinds of glass the less is the corresponding irrationality of dispersion.

The above table is therefore to be supplemented by the following:—

Kind of Glass	*O*152	*S*8	*S*30	*O*60	*O*164	*O*103
α	·628	·645	·644	·643	·637	·605
β	·708	·704	·703	·703	·706	·714
γ	·582	·571	·565	·566	·578	·609

The close agreement between the numbers for *S*8, *S*30, and *O*60 will at once be noticed.

58. Numerical examples of achromatic object-glasses.

The focal length F_D of an objective consisting of two thin lenses close together is calculated for the line D, and achromatism is obtained by choosing the focal lengths f (converging) and f' (diverging) of the component lenses so that the focal lengths of the objective for the lines C and F are equal. Using accented letters to distinguish the constants of the material of which the diverging lens is composed, we have already (Art. 54, equations 109) proved formulae equivalent to

$$\nu f_D = \nu' f'_D = (\nu - \nu')\, F_D \qquad\qquad (120).$$

If squares of $\alpha - \alpha'$, $\beta - \beta'$, $\gamma - \gamma'$ be neglected, it is easy to verify the following additional formulae

$$\left.\begin{aligned} \frac{F_{A'} - F_D}{F_D} = \frac{\alpha - \alpha'}{\nu - \nu'}, \quad \frac{F_C - F_D}{F_D} = \frac{F_F - F_D}{F_D} = \frac{\beta' - \beta}{\nu - \nu'}, \\ \frac{F_{G'} - F_F}{F_D} = \frac{\gamma' - \gamma}{\nu - \nu'} \end{aligned}\right\} \qquad (121).$$

Applying these to a combination of two of the older kinds of glass, *O*60 (converging) and *O*103 (diverging), for which F_F is 1000 mm., we find

$$F_{G'} - F_F = 1{\cdot}79 \text{ mm.}, \quad F_D - F_F = -{\cdot}46 \text{ mm.}, \quad F_{A'} - F_F = 1{\cdot}13 \text{ mm.},$$

numbers which shew clearly the extent of the remaining chromatic defect.

A combination of more modern kinds of glass, *S*30 (converging) and *S*8 (diverging), for which F_F is 1000 mm., gives

$$F_{G'} - F_F = {\cdot}48 \text{ mm.}, \quad F_D - F_F = -{\cdot}07 \text{ mm.}, \quad F_{A'} - F_F = -{\cdot}13 \text{ mm.}$$

These numbers are much smaller than in the previous case, and represent much better achromatism.

The difference between an objective which has been rendered as far as possible achromatic and an object-glass consisting of a single thin lens will be appreciated on comparison of the above figures with those for a converging lens of the glass *O* 60, whose focal length is 1000 mm. A simple calculation gives

$$F_{A'} - F_C = 5{\cdot}8 \text{ mm.}, \quad F_C - F_D = 4{\cdot}9 \text{ mm.},$$
$$F_D - F_F = 11{\cdot}7 \text{ mm.}, \quad F_F - F_{G'} = 9{\cdot}4 \text{ mm.},$$

whence it appears that the foci for the five colours are spread out over a length of 31·8 mm.

IX. THE ABERRATIONS OF THE THIRD ORDER.

59. Improved Approximations.

In the foregoing discussion of the symmetrical optical instrument we have neglected the second and higher powers of the small quantities typified by x, y, δ, ϵ. It is worth while to enquire, without going into any detail, what is the nature of the improved approximation got by neglecting only the fourth and higher powers of these quantities.

At the outset we must notice that in the first order approximation it was legitimate to interpret δ and ϵ as the reduced projected inclinations, or the reduced sines or tangents of the projected inclinations, of the ray to which they referred. In the approximation of higher order it is necessary to be more precise, so δ and ϵ must be understood to be the reduced direction cosines, i.e. direction cosines multiplied by the refractive index of the medium through which the ray is passing. The third direction cosine n is expressed in terms of δ and ϵ by the relation $n = (\mu^2 - \delta^2 - \epsilon^2)^{\frac{1}{2}}/\mu$.

For any selected pair of planes of reference we have obtained approximate formulae of the type

$$\delta = Ax_0 + B\delta_0, \quad x = Cx_0 + D\delta_0.$$

From these, by elimination of δ_0, we derive

$$x = \frac{1}{B}x_0 + \frac{D}{B}\delta \quad \text{.......................(122)},$$

and similarly

$$y = \frac{1}{B}y_0 + \frac{D}{B}\epsilon \quad \text{.......................(123)}.$$

These formulae express the coordinates of the point of intersection of the second plane of reference by that ray from (x_0, y_0) which arrives with the direction defined by (δ, ϵ).

The approximation which takes account of terms up to the third order is got by adding suitable corrections to the above expressions for x and y. It is, in fact,

$$\left.\begin{aligned} x &= \frac{1}{B} x_0 + \frac{D}{B} \delta + f(x_0, y_0, \delta, \epsilon) \\ y &= \frac{1}{B} y_0 + \frac{D}{B} \delta + f(y_0, x_0, \epsilon, \delta) \end{aligned}\right\} \text{..............(124)},$$

where the function f may contain terms of the second and third orders. The symmetry of the instrument is our justification for putting in the expressions for x and y the same function with differently arranged arguments.

From the symmetry of the instrument it is further clear that, to any ray starting from (x_0, y_0) and arriving at (x, y) with direction (δ, ϵ), there corresponds a ray starting from $(-x_0, -y_0)$ and arriving at $(-x, -y)$ with direction $(-\delta, -\epsilon)$. In other words, if the signs of $x_0, y_0, \delta, \epsilon$, be changed, the only change in x and y is a change of sign, and therefore the only change in $f(x_0, y_0, \delta, \epsilon)$ and in $f(y_0, x_0, \epsilon, \delta)$ is change of sign. Hence f cannot contain any terms of the second order; it consists entirely of terms of the third order.

There are 20 terms in the general homogeneous expression of the third order in 4 variables. But from considerations of symmetry we shall shew that many of these cannot be present in $f(x_0, y_0, \delta, \epsilon)$.

In the first place we notice that if we put $y_0 = 0$ and reverse the sign of ϵ, symmetry with respect to the plane xz requires that the value of x shall remain unaltered. Any term which would be altered by this change cannot be present in f. Thus $x_0^2\epsilon$, $\delta^2\epsilon$, ϵ^3, $x_0\delta\epsilon$ must be excluded.

Again, if we put $x_0 = 0$ and reverse the sign of δ, symmetry with respect to the plane of yz demands that the sign of x be reversed. Terms which do not satisfy this condition cannot occur in f. Thus y_0^3, $y_0^2\epsilon$, $y_0\delta^2$, $y_0\epsilon^2$, are to be excluded.

Therefore $f(x_0, y_0, \delta, \epsilon)$ is of the form

$$\begin{aligned} &a x_0^3 + b x_0^2 y_0 + c x_0 y_0^2 + d x_0^2 \delta + e x_0 y_0 \delta + f y_0^2 \delta + g x_0 y_0 \epsilon \\ &\qquad + h\delta^3 + j\delta\epsilon^2 + k x_0 \delta^2 + l x_0 \epsilon^2 + m y_0 \delta\epsilon \text{......(125)}. \end{aligned}$$

Symmetry, however, carries the simplification still further. If we turn the axes of x and y through an angle θ, the new x is $x \cos\theta + y \sin\theta$, the new y is $y \cos\theta - x \sin\theta$, the new δ is $\delta \cos\theta + \epsilon \sin\theta$, and the new ϵ is $\epsilon \cos\theta - \delta \sin\theta$; and x_0, y_0, are changed in the same manner. But the new x must be the same function of the new independent variables as the old x is of x_0, y_0, δ, ϵ. Hence

$$\cos\theta f(x_0, y_0, \delta, \epsilon) + \sin\theta f(y_0, x_0, \epsilon, \delta)$$
$$= f(x_0\cos\theta + y_0\sin\theta,\ y_0\cos\theta - x_0\sin\theta,\ \delta\cos\theta + \epsilon\sin\theta,$$
$$\epsilon\cos\theta - \delta\sin\theta) \quad \text{......(126)},$$

and this must be true for all values of θ.

Differentiate both sides of the equality with respect to θ, and put $\theta = 0$ after differentiation; the result is

$$f(y_0, x_0, \epsilon, \delta) = \left(y_0\frac{\partial}{\partial x_0} - x_0\frac{\partial}{\partial y_0} + \epsilon\frac{\partial}{\partial\delta} - \delta\frac{\partial}{\partial\epsilon}\right) f(x_0, y_0, \delta, \epsilon) \quad \text{......(127)}.$$

On substitution of the twelve-term expression for f in this equality, it is seen that the relation is not satisfied unless the coefficients are subject to the following restrictions:

$$c = a, \quad b = 0, \quad d = g + f,$$
$$e = 0, \quad h = j, \quad k = l + m.$$

So the function f must be of a type involving only six constants, namely

$$f(x_0, y_0, \delta, \epsilon) \equiv ax_0(x_0^2 + y_0^2) + gx_0(x_0\delta + y_0\epsilon)$$
$$+ f\delta(x_0^2 + y_0^2) + h\delta(\delta^2 + \epsilon^2) + lx_0(\delta^2 + \epsilon^2) + m\delta(x_0\delta + y_0\epsilon) \quad \text{...(128)}.$$

It is clear that this form of f satisfies not only condition (127) but also condition (126); for all the expressions in brackets on the right-hand side retain their form when the axes of coordinates are turned through the angle θ, so that f behaves like a linear function of the arguments.

The final simplification is based upon optical considerations, and is got by use of the well-known theorem that there is a function V such that

$$dV = \delta dx + \epsilon dy + \sqrt{\mu^2 - \delta^2 - \epsilon^2}\,dz \quad \text{..............(129)},$$

where V is, in fact, the optical path from (x_0, y_0, z_0) along the ray to (x, y, z). (Cf. Art. 62.)

If we add $d(-\delta x - \epsilon y)$ to both sides we get

$$-x d\delta - y d\epsilon + \sqrt{\mu^2 - \delta^2 - \epsilon^2}\,dz = d(V - \delta x - \epsilon y) \quad \text{......(130)},$$

a total differential, the independent variables now being changed from x, y, z to δ, ϵ, z. From this it follows that

$$\frac{\partial x}{\partial\epsilon} = \frac{\partial y}{\partial\delta} \quad \text{..........................(131)};$$

and consequently that

$$\frac{\partial}{\partial\epsilon} f(x_0, y_0, \delta, \epsilon) = \frac{\partial}{\partial\delta} f(y_0, x_0, \epsilon, \delta) \quad \text{...........(132)}.$$

On substituting in this condition the six-constant formula for f, we find that the equality is not satisfied unless $m = 2l$. Thus the function $f(x_0, y_0, \delta, \epsilon)$ is

$$ax_0(x_0^2 + y_0^2) + gx_0(x_0\delta + y_0\epsilon) + f\delta(x_0^2 + y_0^2) + h\delta(\delta^2 + \epsilon^2) \\ + lx_0(\delta^2 + \epsilon^2) + 2l\delta(x_0\delta + y_0\epsilon) \quad\text{.................. (133)},$$

and involves only five constants.

If we choose the second plane of reference to be the conjugate focal plane of the first plane of reference, D becomes zero, and the formulae expressing x, y in terms of x_0, y_0, δ, ϵ, reduce to

$$\left.\begin{aligned} x &= B^{-1}x_0 + f(x_0, y_0, \delta, \epsilon) \\ y &= B^{-1}y_0 + f(y_0, x_0, \epsilon, \delta) \end{aligned}\right\} \quad\text{............... (134)},$$

where f is of the form of expression (133).

If f were absent, x and y would be independent of δ and ϵ and would be proportional to x_0 and y_0; thus to any bright object in the first plane of reference there would correspond an image similar in form to the object and made up of ideal point-images. The extra terms included in f represent deviations from the ideal image formation and are called aberrations; there are five aberrations, as there are five constants in f.

60. Aberration Curves.

With a view to finding out what changes in the appearance of the image correspond to the various terms in f, it is convenient to introduce new variables instead of δ and ϵ. Let us take another plane, perpendicular to the axis of the instrument, at reduced distance w beyond the second plane of reference, and let (ξ, η) be the coordinates of the point where the ray strikes this plane. Then

$$\delta = (\xi - B^{-1}x_0)/w, \quad \epsilon = (\eta - B^{-1}y_0)/w \quad\text{........ (135)};$$

these results are only approximately true, since they treat δ and ϵ as reduced tangents instead of reduced direction cosines, but they are sufficiently accurate for substitution in f which is small of the third order.

Substituting these expressions for δ and ϵ in $f(x_0, y_0, \delta, \epsilon)$, and for brevity denoting $(Bw)^{-1}$ by λ, we get

$$\{a - (f + g)\lambda + 3l\lambda^2 - h\lambda^3\} x_0(x_0^2 + y_0^2) + B\lambda(g - 4l\lambda + 2h\lambda^2) x_0(x_0\xi + y_0\eta) \\ + B\lambda(f - 2l\lambda + h\lambda^2)\xi(x_0^2 + y_0^2) + B^3\lambda^3 h\xi(\xi^2 + \eta^2) \\ + B^2\lambda^2(l - h\lambda) x_0(\xi^2 + \eta^2) + 2B^2\lambda^2(l - h\lambda)\xi(x_0\xi + y_0\eta) \quad\text{........ (136)}.$$

If, therefore, we introduce a new set of constants we have

$$\left.\begin{aligned} x &= B^{-1}x_0 + \phi(x_0, y_0, \xi, \eta) \\ y &= B^{-1}y_0 + \phi(y_0, x_0, \eta, \xi) \end{aligned}\right\} \quad \text{..............(137)},$$

where

$$\left.\begin{aligned} \phi(x_0, y_0, \xi, \eta) \equiv\ & P\xi(\xi^2 + \eta^2) + Qx_0(x_0^2 + y_0^2) \\ & + R\{x_0(\xi^2 + \eta^2) + 2\xi(x_0\xi + y_0\eta)\} \\ & + (S - T)\,\xi(x_0^2 + y_0^2) + 2Tx_0(x_0\xi + y_0\eta) \end{aligned}\right\} \quad \text{..............(138)}.$$

Since the axes can be turned so as to bring any assigned point of the object into the plane of x, we do not lose any generality by supposing y_0 to be zero. When this is the case our formulae become

$$\left.\begin{aligned} x' &= P\xi(\xi^2 + \eta^2) + Qx_0^3 + Rx_0(3\xi^2 + \eta^2) + (S + T)\,x_0^2\xi \\ y' &= P\eta(\xi^2 + \eta^2) + 2Rx_0\xi\eta + (S - T)\,x_0^2\eta \end{aligned}\right\} \text{......(139)},$$

where x' and y' are written for $x - B^{-1}x_0$ and y respectively, i.e. for the relative coordinates of the point where the ray strikes the plane of the first-order image, relative namely to the point where it would strike this plane if there were no terms but those of the first order.

So far we have not considered the value to be assigned to w. We take the plane specified by w to be the plane of the exit-pupil. This choice has the advantage that it enables us to examine the trace, on the image-plane, of the extreme rays of the pencil, by making the point (ξ, η) traverse the boundary of the exit-pupil, which is of course a circle, say $\xi^2 + \eta^2 = r^2$.

It is, besides, convenient to regard the pencil as made up of subsidiary sheaves or hollow cones of rays which cut the plane of the exit-pupil in circles having their centres on the axis of the instrument. For example a ray of the sheaf which cuts the plane of the exit-pupil in the circle $\xi^2 + \eta^2 = \sigma^2$ may be defined by putting $\xi = \sigma\cos\phi$, $\eta = \sigma\sin\phi$; if we let ϕ take all values from 0 to 2π, (ξ, η) describes its circle, and (x', y') describes in its own plane a curve called an aberration curve. The aggregate of the aberration curves for values of σ varying from zero to r will correspond to the patch of light in the image-plane which takes the place of the ideal point-image.

61. Nature of the five aberrations.

In order to appreciate the physical significance of the five aberration constants, we shall examine in turn the consequences of the presence of each constant when the remaining four are zero.

(i) Suppose P is the only constant which does not vanish. Then

$$\left.\begin{aligned} x' &= P\xi(\xi^2 + \eta^2) = P\sigma^3\cos\phi \\ y' &= P\eta(\xi^2 + \eta^2) = P\sigma^3\sin\phi \end{aligned}\right\} \quad \text{..........(140)}.$$

The aberration curves are concentric circles of radius $P\sigma^3$; and so, instead of a point-image, we have a circular patch of light of radius Pr^3, the size therefore being the same for all points in the object plane. This defect is called *spherical aberration*.

(ii) Suppose Q is the only constant which does not vanish. Then

$$x' = Qx_0^3, \quad y' = 0 \quad \text{......................(141)}.$$

Both x' and y' are absolutely independent of ξ and η, and are therefore the same for all rays proceeding from the object point. So the image of a point is a point. But the distance of the image-point from the axis is not proportional to the distance of the object point from the axis, so the image and the object are not geometrically similar. If Q is positive

$$x/x_0 = B^{-1} + Qx_0^2 \quad \text{......................(142)}$$

and so increases with x_0. If the object is a square with its centre on the axis, the corners of the image-figure will be proportionately further from the axis than the centres of the sides; so the image will be a curvilinear quadrilateral whose sides are concave outwards. If Q were negative the image-figure would have its sides convex outwards.

The defect represented by Q is called *distortion*.

(iii) Let R be the only constant which does not vanish. Then

$$\left.\begin{aligned} x' &= Rx_0(3\xi^2 + \eta^2) = Rx_0\sigma^2(2 + \cos 2\phi) \\ y' &= 2Rx_0\xi\eta \qquad\quad = Rx_0\sigma^2 \sin 2\phi \end{aligned}\right\} \quad \text{......(143)}.$$

The aberration curve corresponding to σ is a circle whose centre has relative coordinates $(2Rx_0\sigma^2, 0)$ and whose radius is $Rx_0\sigma^2$. For different values of σ these circles constitute a series whose centres lie on the line joining the first-order image-point to the axis of the instrument, and whose radii are in each case half the distance of the centre from the first-order image-point. They all touch a pair of straight lines which cut one another at an angle of 60°. Thus the patch of light is wedge-shaped or balloon-shaped, with vertex at the first-order image-point, the wedge extending from or towards the axis according as R is positive or negative.

The defect of the image represented by R is called *coma*.

(iv) The defects represented by the constants S and T are best considered together. Let us suppose that all the constants but these two vanish, and we have

$$x' = (S + T)x_0^2\xi, \quad y' = (S - T)x_0^2\eta \quad \text{...........(144)}.$$

Consider the patch of light which the rays from (x_0, y_0) would make on a screen in the plane parallel to the plane of the first-order image, at a distance ζ further from the instrument. If (α, β) be the coordinates of the point where this plane is met by the ray defined by (ξ, η), it is clear that the part of the ray between the image-plane and the plane of the exit-pupil is divided at (α, β) in the ratio of ζ to $\mu w - \zeta$, so that

$$\mu w\alpha = (\mu w - \zeta)\{B^{-1}x_0 + (S+T)\,x_0^2\xi\} + \zeta\xi,$$
$$\mu w\beta = (\mu w - \zeta)(S-T)\,x_0^2\eta + \zeta\eta.$$

The extreme rays of the pencil are such that $\xi^2 + \eta^2 = r^2$; and the corresponding relation between α and β, namely

$$\frac{\{\mu w\alpha - (\mu w - \zeta)\,B^{-1}x_0\}^2}{\{\zeta + (\mu w - \zeta)(S+T)\,x_0^2\}^2} + \frac{\{\mu w\beta\}^2}{\{\zeta + (\mu w - \zeta)(S-T)\,x_0^2\}^2} = r^2,$$

is the equation to the boundary of the patch of light made on the screen.

The patch is in general elliptical, but one of the principal axes vanishes and the patch becomes a short straight line perpendicular to the plane joining the first-order image-point to the axis of the instrument if ζ has the value ζ_1, where

$$\zeta_1 = -\mu w\,(S+T)\,x_0^2/\{1-(S+T)\,x_0^2\}$$
$$= -\mu w\,(S+T)\,x_0^2 \text{ approximately} \quad\text{.............(145).}$$

This short straight line may be called the primary focal line.

Similarly the patch reduces to a short straight line in the plane through the image-point and the axis of the instrument (the secondary focal line) if ζ has the value ζ_2, where

$$\zeta_2 = -\mu w\,(S-T)\,x_0^2, \text{ approximately} \quad\text{.............(146).}$$

The distance between the two focal lines $|\zeta_2 - \zeta_1|$, is $|2\mu w x_0^2 T|$.

There are two values of ζ for which the patch is circular. One of these is μw which is not necessarily small; but there is one which must be small, namely ζ_3, given by

$$\zeta_3 + (\mu w - \zeta_3)(S+T)\,x_0^2 = -\zeta_3 - (\mu w - \zeta_3)(S-T)\,x_0^2,$$

or

$$\zeta_3 = -\mu w S x_0^2 \text{ approximately} \quad\text{....................(147)}$$
$$= \tfrac{1}{2}(\zeta_1 + \zeta_2).$$

This circular patch is called the circle of least confusion; its radius is $|\mu w T x_0^2|$. If T is different from zero there is no single point through which pass all the rays from (x_0, y_0), that is to say, no point image. The best approximation to an image is represented by the circle of least confusion, and the image picture made up of these circular patches is necessarily slightly blurred. If S is zero, so that ζ_3 is zero, the circles of least confusion corresponding to all the points of the flat object lie in the plane of the first-order image. Thus the best image is flat and is in that plane, but it is made up of overlapping circular patches instead of points. This defect is called *astigmatism*; it is the defect corresponding to the constant T.

If T is zero, while S is not zero, the focal lines and the circle of least confusion coincide and, as the radius of the circle is then zero, they are represented by a single point through which pass all the rays from (x_0, y_0). The pencil is, in fact, stigmatic, there is a point image. But this point image is not in the plane of the first-order image, it lies beyond it at a distance ζ_3 or $-\mu w S x_0^2$, and this distance, moreover, is different for the different points of the object, vanishing only for that point of the object which lies on the axis of the instrument. Thus the image picture of the flat object is stigmatic but curved, being on a surface of revolution which touches the plane of the first-order image at its vertex. If ρ be the radius of curvature of this surface at its vertex, reckoned as positive when the concavity is towards the instrument,

$$\rho = -(B^{-1}x_0)^2/2\zeta_3,$$

whence

$$1/\rho = 2\mu w B^2 S \qquad \text{.......................(148)}.$$

The defect of the image represented by the constant S is called *curvature of the field.*

The manner in which λ enters into the expressions for which P, Q, R, S, T are abbreviations, shews that a proper adjustment of the position of the effective stop may serve to diminish some, or to remove one of the aberrations. But in designing an instrument from which (for a particular image plane) all five third-order aberrations shall be absent, it is necessary to know how the constants depend on the composition of the instrument. The reader who is interested in this subject is recommended to consult Mr R. A. Herman's *Geometrical Optics*, Ch. xiv, or Prof. E. T. Whittaker's Tract on the Theory of the Optical Instrument (Cambridge, 1907). The method of interpreting the aberrations given above is based on Prof. K. Schwarzschild's memoir (*Untersuchungen zur geometrische Optik*, Berlin, 1905); but in the discussion of astigmatism and curvature I have followed a suggestion made by Mr Bromwich, who pointed out that the assumption of the existence of focal lines, legitimate as it is in first-order approximations, requires special justification in an argument which takes account of small quantities of the third order.

X. SYLLABUS OF PROPOSITIONS CONCERNING THE CHARACTERISTIC FUNCTION AND THE FOCAL LINES OF A PENCIL OF RAYS OF LIGHT.

62. Definition and properties of the Characteristic Function.

(i) Let (x_0, y_0, z_0) be the coordinates of a point P_0 situated in a homogeneous medium of index μ_0; let (x, y, z) be the coordinates of a point P, situated in another homogeneous medium of index μ_n; let there be, between P_0 and P, $n-1$ other media whose indices are $\mu_1, \mu_2, \ldots \mu_{n-1}$, separated by refracting surfaces. Consider a path from P_0 to P, made up of straight lines $P_0Q_1, Q_1Q_2, \ldots Q_nP$, the points $Q_1, Q_2, \ldots Q_n$ being on the refracting surfaces. Let the direction cosines of the straight path in the medium μ_r be (l_r, m_r, n_r), and its length s_r, and let the coordinates of Q_r be (x_r, y_r, z_r). Let $\sum_{r=0}^{r=n} \mu_r s_r$ be denoted by W. Denote by δW the difference between W for this path and the corresponding sum for a neighbouring path, $P_0Q_1'Q_2' \ldots Q_n'P'$, of the same type, proceeding from the same starting-point. If the coordinates of Q_r' are $(x_r + \delta x_r, y_r + \delta y_r, z_r + \delta z_r)$, and if all the increments denoted by δ are so small that their squares and products may be neglected, it is seen, by projection, that

$$\delta W = \mu_n (l_n \delta x + m_n \delta y + n_n \delta z) + \sum_{r=1}^{r=n} \{(\mu_{r-1} l_{r-1} - \mu_r l_r)\, \delta x_r + (\mu_{r-1} m_{r-1} - \mu_r m_r)\, \delta y_r + (\mu_{r-1} n_{r-1} - \mu_r n_r)\, \delta z_r\} \quad \ldots\ldots\ldots\ldots (149).$$

(ii) If the original path be the course of a ray of light from P_0 to P, the cosines of the straight pieces which terminate at Q_r will obey the laws of refraction. Consequently the sum which forms the second part of the expression for δW is equal to

$$\sum_{r=1}^{r=n} (\mu_{r-1} \cos \phi_r - \mu_r \cos \phi_r') (L_r \delta x_r + M_r \delta y_r + N_r \delta z_r),$$

the notation being that of Article 1.

Since $(\delta x_r, \delta y_r, \delta z_r)$ are the components of a displacement of Q_r in the surface to which (L_r, M_r, N_r) is normal, each term of this sum is zero.

Consequently the variation of W from an optical path is

$$\delta W = \mu_n (l_n \delta x + m_n \delta y + n_n \delta z) \quad \ldots\ldots\ldots\ldots (150).$$

(iii) For a variation which keeps P fixed, the right-hand side of (150) is zero. Thus, for different paths from P_0 to P, the optical path gives a stationary value of W, usually a minimum. This stationary value is called "the reduced path" or "the optical path" from P_0 to P. It is proportional to the time that light takes to travel from P_0 to P. It is denoted by V.

There is usually only one optical path between two given points. If, however, P_0 and P are conjugate foci, there is an infinite number of optical paths from one to the other, all having (to the degree of approximation contemplated) the same value of V.

(iv) V is a function of (x, y, z) having the properties

$$\frac{\partial V}{\partial x} = \mu_n l_n, \quad \frac{\partial V}{\partial y} = \mu_n m_n, \quad \frac{\partial V}{\partial z} = \mu_n n_n \qquad (151),$$

$$\left(\frac{\partial V}{\partial x}\right)^2 + \left(\frac{\partial V}{\partial y}\right)^2 + \left(\frac{\partial V}{\partial z}\right)^2 = \mu_n^2 \qquad (152).$$

It is called the "characteristic function" of a pencil of rays of light proceeding from P_0.

(v) Rays of light which proceed from a common point-source P_0, possess, after any number of refractions, the property of being cut orthogonally by a family of surfaces. These are the surfaces $V = \text{const.}$ They are the wave surfaces of Physical Optics.

(vi) The preceding propositions can easily be extended to include reflexions.

63*. The modified characteristic function.

(vii) Omitting the suffix n of the previous Article, consider the function U defined by the relation

$$U = V - \mu(lx + my) \qquad (153).$$

In virtue of the equalities

$$\frac{\partial V}{\partial x} = \mu l, \quad \frac{\partial V}{\partial y} = \mu m, \quad n = \sqrt{1 - l^2 - m^2} \qquad (154);$$

it is open to us to regard x, y and n as functions of l, m and z. This is supposed to be done, so that U is a function of the independent variables (l, m, z). And since

$$\begin{aligned} \delta U &= \delta V - \mu\delta(lx + my) \\ &= -\mu(x\delta l + y\delta m) + \mu n\delta z \qquad (155), \end{aligned}$$

it follows that

$$\frac{\partial U}{\partial l} = -\mu x, \quad \frac{\partial U}{\partial m} = -\mu y, \quad \frac{\partial U}{\partial z} = \mu n \qquad (156).$$

U is called the "modified characteristic function."

* I am indebted to Mr Bromwich for the whole of Art. 63.

(viii) Noting that n is a function of l and m only, we may integrate the third of equations (156) partially with respect to z. Thus we see that U must be of the form

$$U = \mu n z + \phi(l, m) \quad\text{.....................(157)},$$

where the form of ϕ depends on the past history of the pencil of rays.

(ix) The foregoing results are exact. If we pass to the case of thin pencils, for which l and m are so small that l^3, $l^2 m$, etc. may be neglected, and if we assume that the origin is on the central ray ($l=0$, $m=0$), so that $\frac{\partial\phi}{\partial l}=0$, $\frac{\partial\phi}{\partial m}=0$, when $l=0$ and $m=0$, we may write

$$\phi = \tfrac{1}{2}(pl^2 + 2qlm + rm^2) \quad\text{..................(158)},$$

where p, q, r are constants defining the form of the pencil.

Thus the approximate form of U for a thin pencil is

$$U = \mu\left[z\{1 - \tfrac{1}{2}(l^2 + m^2)\} + \tfrac{1}{2}(pl^2 + 2qlm + rm^2)\right] \quad\text{...(159)}.$$

(x) This leads to the approximate form of V for a thin pencil, namely by eliminating l and m from

$$\frac{\partial U}{\partial l} = -\mu x, \quad \frac{\partial U}{\partial m} = -\mu y, \quad V = U + \mu(lx + my).$$

The result is

$$V = \mu\left[z - \tfrac{1}{2}\,\frac{(r-z)x^2 - 2qxy + (p-z)y^2}{(p-z)(r-z) - q^2}\right] \quad\text{.........(160)}.$$

(xi) In those optical problems to which the characteristic function can be advantageously applied, the constants p, q, r are not found directly; but the form of V is found for points near to a specified point on the central ray. Let this point be taken as origin, and let the approximate form of V near this point be

$$V = \mu\left[z - \tfrac{1}{2}(ax^2 + 2hxy + by^2)\right].$$

By an elimination analogous to that used in proposition (x), we deduce the corresponding approximation to U, namely

$$U = \mu\left[z + \tfrac{1}{2}(bl^2 - 2hlm + am^2)/(ab - h^2)\right];$$

and a comparison of this with (159) gives the values of p, q, r, namely

$$p/b = q/(-h) = r/a = 1/(ab - h^2).$$

Putting these values in (160) we get

$$V = \mu\left[z - \tfrac{1}{2}\,\frac{ax^2 + 2hxy + by^2 - (ab - h^2)z(x^2 + y^2)}{1 - (a+b)z + (ab - h^2)z^2}\right] \quad\text{...(161)},$$

for points near the axis of z but not necessarily near the origin.

(xii) The formulae of the last three propositions are greatly simplified if the axes are so chosen that h (and consequently q) is zero. In this case, putting

$$p = 1/a = \rho_1, \quad r = 1/b = \rho_2,$$

we get the standard forms

$$V = \mu\left[z - \tfrac{1}{2}\left(\frac{x^2}{\rho_1 - z} + \frac{y^2}{\rho_2 - z}\right)\right] \quad \text{..................(162)},$$

$$U = \mu\left[z + \tfrac{1}{2}\{(\rho_1 - z)\,l^2 + (\rho_2 - z)\,m^2\}\right] \quad \text{...........(163)}.$$

(xiii) V depends on x_0, y_0, z_0, in such a way that

$$\frac{\partial V}{\partial x_0} = -\mu_0 l_0, \quad \frac{\partial V}{\partial y_0} = -\mu_0 m_0, \quad \frac{\partial V}{\partial z_0} = -\mu_0 n_0 \quad \text{.........(164)}.$$

U depends on x_0, y_0, z_0, in such a way that

$$\frac{\partial U}{\partial x_0} = -\mu_0 l_0, \quad \frac{\partial U}{\partial y_0} = -\mu_0 m_0, \quad \frac{\partial U}{\partial z_0} = -\mu_0 n_0 \quad \text{........(165)}.$$

(xiv) Let a pencil of rays set out from a point (x_0, y_0, z_0) very near the axis of a symmetrical optical instrument and traverse the instrument; and let it be desired to find the form of V and U for the emergent pencil.

Take the plane $z = z_0$ for the first plane of reference, and the plane $z = 0$ for the second plane of reference. Let the corresponding constants be A, B', C', D'. It will be found that, for points very near to the origin,

$$V = \mu z - \mu_0 z_0 + \frac{1}{2D'}\{B'(x^2 + y^2) - 2(x_0 x + y_0 y) + C'(x_0^2 + y_0^2)\} + \text{const.} \quad \text{.........(166)},$$

$$U = \mu z - \mu_0 z_0 + \frac{1}{2B'}\{A(x_0^2 + y_0^2) - 2\mu(l x_0 + m y_0) - \mu^2 D'(l^2 + m^2)\} + \text{const.} \quad \text{.........(167)}.$$

For points which, though near the axis, are not very near the origin, the corresponding forms are got by substituting $C' + A\mu^{-1}z$ for C', and $D' + B'\mu^{-1}z$ for D'.

64. The focal lines of a pencil of rays.

(xv) The normals to a surface at points in the immediate neighbourhood of an ordinary point of the surface pass approximately through two straight lines at right angles. These lines are parallel to the tangents to the lines of curvature at the point considered, and pass through the centres of principal curvature.

To indicate the degree of approximation, let (ξ, η, ζ) be a point near the origin, on the surface whose approximate equation near the origin is $2z = x^2\rho_1^{-1} + y^2\rho_2^{-1}$. The shortest distance between the normal at (ξ, η, ζ) and the line $x = 0$, $z = \rho_1$, is $\xi\zeta\rho_1^{-1}$ plus terms of higher order. The shortest distance between the normal and the line $y = 0$, $z = \rho_2$, is $\eta\zeta\rho_2^{-1}$ plus terms of higher order. These shortest distances are small of the third order in ξ and η.

(xvi) A thin pencil of rays, proceeding from a point-source of light, will, after any number of refractions or reflexions, pass approximately through a pair of focal lines at right angles, namely the focal lines of any one of the surfaces to which the rays are normals.

(xvii) Normals are drawn to the surface $2z = x^2\rho_1^{-1} + y^2\rho_2^{-1}$ at the points where it is met by the thin elliptic cylinder $x^2/\alpha^2 + y^2/\beta^2 = 1$. The trace of these normals on the plane $z = p$ is approximately the ellipse

$$x^2/\alpha^2(1 - p\rho_1^{-1})^2 + y^2/\beta^2(1 - p\rho_2^{-1})^2 = 1 \qquad (168).$$

The changes in the size and shape of this ellipse, as p varies, should be studied.

For a particular value of p the ellipse is a very small circle, namely of radius

$$(\rho_1 \sim \rho_2)/(\rho_1\alpha^{-1} + \rho_2\beta^{-1}) \qquad (169).$$

In the optical application this is the circle of least confusion.

It frequently happens that the pencil of rays is bounded by a circular stop or the circular aperture of a lens, through which the pencil passes obliquely, the angle of incidence being ϕ; and therefore frequently $\alpha/\beta = \cos\phi$.

(xviii) The oblique refraction of a thin pencil of rays is considered in the particular case specified by the following limitations: (1) The focal lines of the incident pencil are supposed to be respectively in and perpendicular to the plane of incidence of the central ray of the pencil; (2) the focal lines of the normals to the refracting surface are supposed to be respectively in and perpendicular to the plane of incidence.

When terms of a higher order of smallness than the second are neglected in the equations of the refracting surface and the orthotomic surfaces of the pencil, these suppositions guarantee symmetry about the plane of incidence. Hence the refracted pencil also has its focal lines in and perpendicular to the plane of incidence.

Focal lines perpendicular to the plane of incidence are called "primary." P_1, the point where the primary focal line of the incident pencil cuts the plane of incidence, is the limiting position of the point of intersection of the central ray of the pencil with a consecutive ray which is also in the plane of incidence. Q_1, the corresponding point for the refracted ray, is defined in a similar manner. O_1, the point where the primary focal line of the refracting surface meets the plane of incidence, may be called the primary centre of curvature; it is the limiting position of the intersection of consecutive normals to the surface, both in the plane of incidence, one being at the point of incidence of the central ray.

The secondary focal lines are marked by their intersections P_2, Q_2, O_2 with the central incident ray, the central refracted ray, and the normal to the refracting surface at the point of incidence K of the central ray. They are the limiting positions of the points of intersection of these lines with the corresponding lines for a consecutive point of incidence K', such that the arc KK' is perpendicular to the plane of incidence.

Distances being measured positively into the second medium, the position of Q_1 is determined by the equation,

$$\frac{\mu' \cos^2\phi'}{KQ_1} - \frac{\mu \cos^2\phi}{KP_1} = \frac{\mu' \cos\phi' - \mu \cos\phi}{KO_1} \quad \text{.........(170)}.$$

By the first law of refraction it is seen that P_2, Q_2, O_2 lie in a straight line, and so the position of Q_2 is determined by the equation,

$$\frac{\mu'}{KQ_2} - \frac{\mu}{KP_2} = \frac{\mu' \cos\phi' - \mu \cos\phi}{KO_2} \quad \text{..............(171)}.$$

This theorem is of wider application than might at first be supposed. Restriction (2) is satisfied for a spherical surface, whatever be the plane of incidence. Restriction (1) is satisfied if the incident pencil is homocentric. In many applications (1) is satisfied in virtue of the previous history of the pencil; for example if a homocentric pencil enters a symmetrical optical instrument, and if its central ray intersects the axis of the instrument, then at every refracting surface in the instrument the incident pencil will satisfy this restriction.

(xix) The formulae for reflexion which correspond to (170) and (171) may be derived from these by putting $\mu' = -\mu$. It is easy to prove them independently.

65. Alternative methods of finding the form of V.

(xx) Choose origin and axes so that one of the orthotomic surfaces of the pencil under consideration is $2z = x^2/\rho_1 + y^2/\rho_2$; express as a function of the coordinates (x, y, z) of a point P, the length of the normal from P to the surface. This, multiplied by μ, must differ from the value of V at P only by a constant. Thus formula (162) is obtained.

(xxi) Another method of proof is to assume that V can be expanded in the form

$$V = A + Bx + Cy + Ex^2 + Fxy + Gy^2,$$

where the great letters represent functions of z. The axis of z being a ray of the pencil, $\partial V/\partial x$ and $\partial V/\partial y$ must vanish if $x = 0$ and $y = 0$, and $\partial V/\partial z = \mu$; hence $B = 0$, $C = 0$, $A = \mu z + \text{const.}$ By suitable choice of coordinate planes F can be got rid of. The functions represented by the other letters must be such as to make the equation

$$\left(\frac{\partial V}{\partial x}\right)^2 + \left(\frac{\partial V}{\partial y}\right)^2 + \left(\frac{\partial V}{\partial z}\right)^2 = \mu^2$$

correct to the second order in x and y. Substitution gives differential equations with respect to z, whence the forms of E and G can be determined. The constants ρ_1, ρ_2 are constants of integration.

(xxii) The transition from (162) to (161) is effected by means of the following theorem in two-dimensional analytical geometry :—

The orthogonal transformation of axes which converts $X^2/\rho_1 + Y^2/\rho_2$ into $ax^2 + 2hxy + by^2$, will convert

$$X^2/(\rho_1 - \lambda) + Y^2/(\rho_2 - \lambda)$$

into
$$\frac{ax^2 + 2hxy + by^2 - \lambda(ab - h^2)(x^2 + y^2)}{1 - (a + b)\lambda + (ab - h^2)\lambda^2}.$$

(xxiii) The focal lines of the pencil whose characteristic function is given by (162) are $z = \rho_1$, $x = 0$, and $z = \rho_2$, $y = 0$.

66. The general problem of oblique refraction.

(xxiv) From its definition as a reduced path, it is clear that the characteristic function cannot undergo any abrupt change of numerical value, as the pencil traverses a refracting surface. But the characteristic functions of the incident and the refracted pencils are different.

The general problem of the oblique or normal refraction of a given thin pencil at a given refracting surface can be solved by obtaining the characteristic function of the refracted pencil; this function is determined by its having to satisfy two conditions, namely that it shall be of the standard algebraical form when referred to the standard axes of the pencil, and that it shall be arithmetically continuous, at the refracting surface, with the known characteristic function of the incident pencil.

Normal refraction through an astigmatic lens (e.g. a cylindrical lens) is easily worked out in this manner.

67. Heterogeneous media.

(xxv) A ray in a heterogeneous medium (μ a function of x, y, z) satisfies the differential equations of the type

$$\frac{d}{ds}\left(\mu \frac{dx}{ds}\right) - \frac{\partial \mu}{\partial x} = 0 \quad \text{.....................(172)},$$

where ds represents an element of arc on the ray. There are three such equations, only two of them, however, being independent. They may be obtained as limiting forms of equations (1).

(xxvi) Let (x_0, y_0, z_0) be the coordinates of a point P_0 situated in a medium, not necessarily homogeneous, whose index is denoted by μ_0. Let (x, y, z) be the coordinates of a point P, situated in another medium whose index is μ_n. Let there be, between P_0 and P, $n-1$ other media, $\mu_1, \mu_2, \ldots \mu_{n-1}$, separated by refracting surfaces (i.e. surfaces where there is abrupt change of the index). Let a curve, liable to abrupt changes of direction at the refracting surfaces but otherwise continuous, be drawn from P_0 to P, cutting the refracting surface (μ_{r-1}, μ_r) in a point Q_r whose coordinates are denoted by (x_r, y_r, z_r), and the other refracting surfaces in points specified by suitable changes of suffix.

Let $\int_{P_0}^{P} \mu \, ds$, taken along this curve from P_0 to P, be denoted by W, and let the path be varied in the manner contemplated in the ordinary theory of Calculus of Variations, P_0 being kept fixed, but P being allowed to vary. Then

$$\delta W = \left\{\int_{P_0}^{Q_1} + \int_{Q_1}^{Q_2} + \ldots + \int_{Q_n}^{P}\right\} \left\{\left[\frac{\partial \mu}{\partial x} - \frac{d}{ds}\left(\mu \frac{dx}{ds}\right)\right]\delta x + \ldots\right\} ds$$

$$+ \mu_n \{l \delta x + m \delta y + n \delta z\} + \sum_{r=1}^{r=n} \{(\mu_{r-1} l_{r-1} - \mu_r l_r') \delta x_r$$

$$+ (\mu_{r-1} m_{r-1} - \mu_r m_r') \delta y_r + (\mu_{r-1} n_{r-1} - \mu_r n_r') \delta z_r\},$$

where (l, m, n) are the direction cosines of the tangent to the path at P, (l_r', m_r', n_r') the cosines of the tangent to the path at Q_r in the medium μ_r, and $(l_{r-1}, m_{r-1}, n_{r-1})$ the cosines of the tangent to the path at Q_r in the medium μ_{r-1}.

(xxvii) If the original curve be, in each medium, a ray of light, the first term in the expression for δW vanishes in virtue of the differential equations satisfied by a ray.

(xxviii) If the changes of direction of the original curve at the refracting surfaces obey the laws of refraction for optical rays, then the third term in the expression for δW is equal to

$$\sum_{r=1}^{r=n} (\mu_{r-1} \cos \phi_r - \mu_r \cos \phi_r') (L_r \delta x_r + M_r \delta y_r + N_r \delta z_r),$$

the notation being that of Art. 1; and, since $(\delta x_r, \delta y_r, \delta z_r)$ is a displacement of Q_r in the surface to which (L_r, M_r, N_r) is normal, each term of this sum is zero.

(xxix) If the original curve be the course of a ray of light from P_0 to P, and if both P_0 and P be fixed, $\delta W = 0$. For different paths from P_0 to P, the optical path gives a stationary (usually a minimum) value of W. This stationary value is called "the reduced path" or "the optical path" from P_0 to P. It is denoted by V.

(xxx) V thus defined satisfies (151) and (152), and proposition (v) is valid for a pencil which has traversed or is traversing a heterogeneous medium.

Printed in the United States
By Bookmasters